Making Sense of Making Scents

Making Sense of Making Scents

Essential oils and the sweet smell of success

Bruce Dolby

authorHOUSE®

AuthorHouse™
1663 Liberty Drive
Bloomington, IN 47403
www.authorhouse.com
Phone: 1-800-839-8640

© 2011 by Bruce Dolby. All rights reserved.

No part of this book may be reproduced, stored in a retrieval system, or transmitted by any means without the written permission of the author.

First published by AuthorHouse   06/13/2011

ISBN: 978-1-4567-7944-3 (sc)

Printed in the United States of America

Any people depicted in stock imagery provided by Thinkstock are models, and such images are being used for illustrative purposes only.
Certain stock imagery © Thinkstock.

Because of the dynamic nature of the Internet, any web addresses or links contained in this book may have changed since publication and may no longer be valid. The views expressed in this work are solely those of the author and do not necessarily reflect the views of the publisher, and the publisher hereby disclaims any responsibility for them.

Contents

Introduction: Essentials that make Scents........................vii

Chapter 1: A Brief(ish) History 1

Chapter 2: The Marketplace Today 17

Chapter 3: Janet and John play at Science 25

Chapter 4: Janet and John make a Smell........................ 37

Chapter 5: Testing, Testing—Having a Brew 49

Chapter 6: Doing it for Real.. 59

Chapter 7: Now What? 69

Chapter 8: What Now? 77

Chapter 9: Just be Careful out there 85

Chapter 10: And so to Bed 95

Compulsory Thanks.. 103

Introduction

Essentials that make Scents

When I was first asked if I could prepare a small book along the lines of 'Janet and John go distilling' I was both flattered and flattened: the former because I have never written a book before—and the latter because . . . I have never written a book before. Ah, you've guessed . . . Anyway, as both Shakespeare and Mickey Spillane had to actually start somewhere, this seems as good a place as any for The Great Initiation, Book 1, part 1. Introduction. Start. Now!

To date I had only written light-hearted/inconsequential/puerile articles in the Smallholder magazine and the like, including one outlining the fact that there has never been a better time for farmers to grow plants for essential oils and preferably to be able to distil the essential oils on the farm. It certainly seems preferable to getting 26 pence a litre for your milk when it costs 28 pence to produce it, or allowing oiks on quad bikes to go careering through your woodland shooting paintballs at families picnicking along your Nature Trails (or whatever the latest fad is . . .) And whilst the thought of whiskery coach trippers pouring into Marigold's Gift Shoppe or Rosemary's Tea Room is enough to make the average farmer wake up with palpitations: if the bottom line is black rather than red it should be an option at least worthy of a second look. At least it is farming—of a sort anyway and is surely preferable to yet another blasted golf course.

Everyone, and especially the Agricultural press, says that farming today is in crisis and is going directly to hell in a handcart. The problem is basically this: all the time agricultural products can be cheaply flown in from around the globe, from places that are either climatically or politically advantaged and at prices that do not reflect our own wages and overheads, then farmers here aren't even in with a shout.

Supermarkets call all the shots and price is king. So it seems likely that if we are not careful, farming in Britain will go the same way as mining, shipbuilding and fishing, i.e. it will simply become an irrelevance, an extinct profession found only in history books.

The difference is this: the majority of people who live on these islands (John Prescott aside) rather like farming the way it is. They like the farms and the barns, the hills and dales, the streams and the duck ponds, the pretty green bits and even the odd brown stinky bits. In addition, they feel slightly uneasy buying Kenyan runner beans in December or apples that have been flown thousands of miles and then sold for a few pence each.

Hear Ye, Oh Farmers! Deep in their hearts, local people are gunning for you!

It is therefore up to the farmers themselves to produce something that snags the conscience of the local population.

Making Sense of Making Scents

Just look at the success of the Farmers Market phenomenon, where people are happy to pay twice the price just for spuds with dirt on or eggs with worse. So my mission is to boldly go where no adviser has boldly gone before and write down the benefits of growing plants for Essential Oils—and the joy of being able to distil the oil from them—on the farm.

As for the housing; every farm has at least one disused barn—you know, the one round the back that's full of old sacks, a seized up muck spreader and a 1968 Royal Enfield motorbike with the back wheel missing. I know you were thinking of trying to get planning permission to turn it into a holiday home, but what about putting a Still in there and then letting it earn you considerable sums of money? There are other advantages too—for the first time since its construction your barn will not smell like an African tannery on an off day and you won't have to put up with holiday-makers leaving in a huff because 'their daughter keeps treading in all the do-dos everywhere'. Your farm will become a talking point and if you let it, a focal point for visitors too: these visitors may have come from many miles away and are itching to part with some of their folding stuff on souvenirs or fragrant presents for sisters and girlfriends and bedridden Aunties. What better than a range of Essential Oils, or soaps, bath oils and other cosmetic smellies all 'containing genuine Lavender Oil, grown and distilled on the premises here, in deepest Darwentshire'.

'The boxed set of all six soaps is on special offer of just £14.95'. Kerching!!

Bruce Dolby

'Our 'Aromatherapy Selection' contains four different essential oils for £17.75, or available singly for only £5 each'. Kerching!!

'Yes, the farm shop is just opposite and they should have organic apples'. Kerching!!

Naturally I don't expect essential oil distilleries to spring up across the country like mushrooms, (or should that be wind farms?), but there is certainly room for many, many more. Indeed, in Victorian times there were numerous farms growing Lavender, Peppermint, Rosemary and dozens of other aromatic plants. In many cases the essential oils were produced in ancient copper stills dragged into the fields where the plants grew, with heat provided from burning faggots underneath. Whilst they have all disappeared from the British countryside (apart from the faggots), the demand for essential oils most certainly hasn't—and is still growing.

For a great many smaller farms, the installation of a modern, gleaming stainless steel still is certainly worth investigating, and it must surely represent a better bet than a lot of the more traditional current options available, which leave barely enough profit to pay for a call to the Samaritans.

Hopefully this little book will instil some ideas for farming in the future—and if it inspires any farmer to give essential oils a try then I would consider it to have been a success. And for those who simply cannot envisage such an outrageous diversion from traditional farming practices, then I suppose there is always the Paintball option.

Chapter 1

A Brief(ish) History

Smelly people have been around for a long time. I can't prove it of course, but it's a pretty safe bet to assume that B.O. goes way back into B.C. and then a fair bit more. Clever anthropologists are now able to deduce from a chipped molar and a dollop of resinous pollen that the Neanderthal man in question died in his 40's in the middle of March and had suffered from rickets and an early form of bird flu: so how long will it be before they can say with certainty that prehistoric Lucy had halitosis? Or that the Peat Bog Man was actually strangled by his wife who could no longer stand his revolting feet? After all, they have still got the remains of his boots to look at.

Where there's a whiff there's a way—and the only way to overcome the foul and fetid was to use a nice niff as a counter—i.e. pleasant perfumes to obliterate the sulphurous stench. Hence, the evolution of man—leading inevitably and immediately to the production of essential oils and perfumes.

I stink, therefore I am.

Bruce Dolby

Yes, we've been making fragrances for millennia since the burning of incense in Egyptian times: a quick leap forward to the biblical era and I can now imagine Mary being really pleased to get the frankincense and myrrh, what with the cows parping and splattering and all that methane floating about, as well as a bunch of smelly shepherds traipsing in without wiping their feet. Fast-forward a few centuries more and now there are ladies breathing through posies to avoid the noxious vapours, and so it continues through all that nosegay nonsense right up to today's Glade decade.

To be fair, this little book is hardly the place to give an exhaustive treatise on perfumery from 6000 B.C. onwards. In fact I can't think of better way to induce the kind of deep sleep that leaves you waking with dribble running down your chin than to offer a comprehensive history on the use of aromatic products. For goodness sake, that is exactly why the Internet was invented. If you need proof, gently click your mouse (or mouse equivalent) to 'Google' and type in 'Essential oil history': I think you will find approximately 12.8 **million** web sites spring forth, with most of them bursting a blood vessel to tell you all about it; chapter, verse and worse. O.K., some of the later ones might have a slightly tenuous link between Essential oil history and their own, somewhat surprising subject matter, but that still leaves an incredible amount of factual mind-numbing information guaranteed to leave you glassy-eyed and gibbering by the time you throw in the towel and crawl off to bed. (For the remaining 782 people in Britain who haven't the faintest idea what this Internet stuff is all about, I can only apologise).

What I **would** like to spotlight is that until relatively recently—say in the period up to the second world war, there were essential oils being produced locally and often in the field where the crops were grown. In France and Spain this practice was widespread and it still exists there today in remote areas, as indeed it does in the nooks and crannies of other Mediterranean countries too. Let me say right from the beginning, **No**, we are **not** talking poteen or any other form of bootleg hooch here: the stills may look like they were hidden away from prying eyes, but the chances are they were simply lugged into the woodland at the edge of the fields because that was where the firewood was collected. (Of course you always *could* distil something quietly when no one was around but the output would hardly be drinkable and would certainly present an interesting challenge to the wine aficionados. They are already way over-the-top with their descriptions e.g. 'Powerful, complex, flowery notes with exceptional length and gooseberry overtones.' I suppose this would then have to be replaced by something like 'Devastating left hook between the nostrils with a turpentine afterburn').

Figure 1

Figure 1 shows a typical one-man-band distillery, although in this photograph Jill accompanies Jacques; both of them demonstrating a certain chic that is, how you say, oh so typically French. This is an old Eucalyptus still that was indeed operating in the south of France and photographed before the First World War (they have just been caught having a brew).

Figure 2

Figure 2 shows an ancient Lavender still, again used in the South of France where these little stills were commonly seen boiling and blooping away during the late summer and early autumn months. In the more mountainous areas of the Alps and Pyrenees they still carried out distillations, but of plants that were more suited to growing at that altitude, such as juniper (berries) and varieties of pine etc. - see *Figure 3*.

(No, that's not the Gendarme coming to check it out for moonshine production or perhaps a petit Pernod: the owner just happens to enjoy dressing up in his Batman outfit).

Making Sense of Making Scents

Figure 3

The localised production of essential oils was not always performed on such a tiny scale either; in *Figure 4* we can see a far more commercial production in operation even though the picture was taken in the first years of the 20th Century. This shows the production of Spike lavender oil in Spain with the crop being brought to the still by donkey power. Manpower never seemed to be problem in those days either, with plenty of rural hombres available, even if they don't appear to be suffering an adrenaline rush.

Bruce Dolby

Figure 4

Another option was to use a traction-engine to provide the steam; after all, at that time these great lumbering machines were chuntering all around Europe, looking for things to do to keep them busy and out of mischief. They threshed, pumped, hauled, sawed, hammered and here one is even providing the steam to make essential oils *Figure 5*. This one is rather delightfully known as 'an itinerant still', and was producing Lavender oil in the Basses-Alpes region of France in around 1915 when this photograph was taken. I would imagine that if these stills were operating today, then the lads from Health and Safety might possibly like to have a quiet word with the boss (possibly the Groucho Marx impersonator on the right).

Making Sense of Making Scents

Figure 5

Figure 6

With time the tiny producers faded: essential oils were still produced on the farms, but the production methods usually became larger and slightly more sophisticated. The commercial oil distilleries that were run for the fragrance houses in the towns were now producing oils in relatively vast quantities: if the farms wanted to be able to supply them, then their own output had to increase accordingly. *Figure 6* shows a relatively modern set-up: this was producing Rosemary oil on the Island of Lesina (now known as Hvar in the Adriatic Sea and part of Croatia). It still needed plenty of operatives, but of course in those times, labour was as cheap as chips.

Figure 7

The still I used to operate as a teenager in the late 1960's looked only slightly more up to date than this drawing of a 'modern distillation plant' circa 1926, (see *Figure 7*). I remember that my one was certainly a lot filthier—in fact it

Making Sense of Making Scents

was hard to see where the gunge ended and the still began. Steam used to occasionally blast out and frazzle your legs and nether regions as well if you got too close, but at least our clock was identical to the one in the picture (except that ours never worked).

Figure 8

Bruce Dolby

Figure 9

No matter how efficient all these distilleries were, if they didn't have the necessary supply of plant materials they would be as much use as a Ferrari without the fuel. This is where the backbreaking part of the operation comes into play and fortunately where enormous advances have now been made. Today there are specialised harvesters designed to tackle all manner of plants, shrubs, leaves and roots and we will surely never again see those great gangs of rural labourers working tirelessly in the fields (except, of course, during 'Pick-your-Own' at strawberry time).

Figure 10

Figure 8 is copied from an ancient photograph of ladies working in the Jasmine fields near Seillans in the Department of Var in southern France at the turn of the century—and I don't mean this one. Meanwhile, back in Blighty, our

Making Sense of Making Scents

own gels weren't simply sitting back painting their nails and doing lunch—here is a picture (*Figure 9*) of a bevy of our finest English womanhood enjoying themselves cutting Lavender in the fields of Fring in Norfolk, all destined for the Yardley scent factory. Whilst our girls can't compete in the hat stakes they can certainly shift some pretty weighty tote-bins, compared to the somewhat wimpy baskets carried by their French counterparts.

Figure 11

Figure 10 shows another British operation: this is carting Peppermint to the stills in Long Melford, Suffolk, whilst *Figure 11* shows the men doing some work for once. Here they are cutting Peppermint in Michigan, USA and I would imagine that if you were left-handed, you would have stood absolutely no chance of a job with this team.

Bruce Dolby

Figure 12

The Peppermint was taken to the still nearby and you can see that once again labour didn't pose too much of a problem (see *Figure 12*). However for real manpower, (although in this case it seems to be 99% girl power), here is a photograph from the 1920's taken at the Lautier Fils factory in Grasse (*Figure 13*). The only male on the scene is the foreman who is leaning against the wall, (typical!) whilst the other 30+ workers are all women. They are making pomades and all of them look pretty hacked off: presumably it must be a hateful job on a par with emptying bedpans—only slightly more fragrant (or maybe they are simply fed up with their idle boss).

Figure 13

Not that many years ago, certain bits of the British Isles were famous for their own 'special' essential oil. We have already seen the harvesting of Norfolk Lavender: a firm favourite with ladies 'blessed with a certain degree of experience and seniority'. Another English oil was Peppermint, synonymous with Mitcham in Surrey and it was, of course, the best in the world . . . (yeah, yeah). There are still Lavender fields in Norfolk, which produce glorious oils, but compared to the acreage (or should that be hectarage?) grown and the oil production achieved in France for example, then we are comparing the equivalent of Thomas the Tank Engine to a TGV. As for Mitcham and its fabled mint, well, you can forget it! The only chance you've got of finding anything whatsoever growing in this particular concrete landscape is perhaps stumbling across a backyard greenhouse chock full of Category C, (for private consumption only). Mitcham Mint was famous around the world and now only the aroma lingers on. The distillation plant closed over a hundred years ago, although the mint fields did struggle on for several decades more until the builders and developers inevitably

Bruce Dolby

had their wicked way. There are now very few essential oil distilleries left in Britain and the ones that do remain are nearly all pretty hefty industrial affairs with perhaps just a few smaller units.

It wasn't always like that. I am currently the proud borrower of a delightful booklet entitled 'The Still Room', written by a certain Mrs. Charles Roundell and dated MDCCCCIII (or 1903 in decimal). As you can probably guess, it is written in a style that would probably not go down too well with Virago Press. The first chapter is entitled 'A Plea for Housewifery' and contains absolute gems of its time, such as "If it is not beneath the dignity of a man to spend enjoyable hours of labour in laboratory or malthouse, woman need not feel degraded by the apportionment to her of those duties which are more immediately bound up with the creation of happy and refreshing homes." Today this approximates to 'I'll just pop down the pub, love, while you vacuum the lounge'.

There then follows chapters on Butter and Cream, Cheese, Pickling Meat, Fish, Preserves, etc. etc. with copious recipes. (I particularly liked the 'Toast and Water' in the 'Food for Invalids' section; this comprises a thin slice of toast covered with a quart of freshly drawn cold water: allow to stand for one hour, remove toast to produce a refreshing and nutritious drink . . . mmmmmm!)

However, three quarters of the way through the book and Bingo!—a chapter entitled "The Distilling of Waters and Cordials". It starts as follows: "There is no occupation that comes nearer to the work of gods than this occupation of distilling. By the application of fire, the purest of the elements, we separate from gross, substantial bodies those subtle essences which alone give them distinction and charm . . ." It then goes on to describe how to distil your own essential oils for making your own perfumes, but then rather blots its copybook by giving recipes for Absinthe and how to distil liqueurs at home. It also has some fascinating photographs of the stillroom of the London Essence Company, dated 1903, which look remarkably similar to where I used to work.

When I started my first proper job in the late 1960's, it was in a small factory in Kent which produced perfumes and flavours. In those days we actually manufactured the essential oils as well as oleoresins and other perfumery goodies and at that time there were lots of small companies producing these oils. But even then, all of the raw materials arrived from abroad—Buchu leaves from South Africa, Ginger from Zanzibar, Oakmoss from Yugoslavia, Orrisroot from Lord knows where . . . I don't think that **anything** was of U.K. origin. This is a great shame since this country could have produced so much . . . and it still could today.

With modern technology coupled with most of the population's increased holiday/leisure/where-shall-we-go-this-Sunday? time, there is a real possibility of a resurgence of small-scale essential oil distillation. This is particularly true in those areas of Britain that are fortunate enough to benefit from geographic advantages e.g. sited in an area with an established tourism infrastructure, or with particularly good soil or climatic conditions, or having easy access to local marketplaces etc.

Small-scale commercial essential oil production could, and indeed **should,** come back into existence in Britain: only this time the crop would be taken to a local still rather than being primitively distilled in the fields. There is now no compulsion to wear daft hats whilst harvesting and the donkeys can at last be allowed to retire to Blackpool. The numerous sidelines and spin-offs would also add profitability to the venture and it is not unreasonable to expect that these enterprises would provide a lifeline to many besieged British farmers and smallholders. It is a venture that is as environmentally friendly as it is possible to be, great for the local wildlife (butterflies in particular will think they have died and gone to heaven) and it would also help the local economy.

But most of all it would be **fun**. Surely nothing can beat the hypnotic effect of watching those beautiful golden drops of fragrant oil splashing down and rising gently to the surface to form a beautiful bank-balance-in-the-black layer. Lovely!

Chapter 2

The Marketplace Today

Before anyone even thinks about jumping on to this essential oil scentwagon, it pays to look at the marketplace today. What is the profitability likely to be? What are the long-term prospects, if any? Only foolish virgins would pile in willy-nilly, so let's give all the pros the once-over as well as tackling the cons.

Unlike the farming co-operatives and societies that operate so brilliantly in France and to a lesser extent in Britain too, there is no single body that represents and fights for the interests of small essential oil growers or distillers. Who knows, one day there may be a "Society for Oil Distillation Suppliers" for example, (although a different acronym may be preferred) but at the moment, for anyone interested in producing essential oils and their products then I'm afraid that, society and assistance-wise, you're on your own chum. On the positive side at least you wouldn't have to suffer all that committee baloney, with its accompanying memos, election of officers, quorums, forums, members enquiries, letter from abroad and the predictable 'I'm-afraid-there-will-have-to-be-an-increase-in-subscriptions-next-quarter' announcement. Finally, you just know that one day something will happen that will tip you over the edge and guarantee that you leave this bunch of saddos. For "The committee has arranged for an expert to give a talk on *exciting new developments* taking place in the field of essential oil distilling . . ." read "A wet behind the ears ignoramus from an obscure government quango will pontificate and attempt to dispense advice in a village hall on something he's been asked to mug up on the day before" . . .

Relatively speaking, there are very few small-scale farmers of any shape or form actually *left* operating in Britain, so the number that are currently growing crops for essential oils is next to negligible. What there **is** an absolute mountain of (and we're now talking numbers big enough to make cheapo calculators display a row full of EEEEs) is the amount of enterprises flogging every essential oil under the sun, in bottles of every imaginable shape and size right down to miniscule 2ml vials containing less than half a teaspoonful. And what about all the outfits who use essential oils in their products? In fact it's probably easier to count the producers who don't market products that use or contain essential oils. For a start just look at all that aromatherapy stuff on sale in every town and village throughout the country. Then think about the soaps, bath oils, body sprays, hand creams, nail gels, floral waters, shampoos, foot lotions, candles, pot pourri and every other fragranced goody you can imagine as well as some you wouldn't want to think about?

What is so sad is that nearly all of the oils are distilled from plant material that was grown abroad with only a tiny percentage coming from Britain. Take a look on the Internet yourself and you will see France, Portugal, Somalia, Brazil and Sri Lanka are doing well, but Madagascar, the Comoros and India are doing better. Egypt and Morocco are in joint

Making Sense of Making Scents

second but the winner by about 2 lengths is China. The oils that **are** produced in Britain are normally distilled in relatively large factories on a considerable industrial scale, supplied by pretty hefty farms. These factories may be extremely efficient—and to be fair, capable of producing superb oils—but they certainly would not have any bearing or effect on the viability of your own enterprise.

There are several major players in the essential oils game and they often use these oils as components of the perfumes that then go in branded products such as 'high street' soaps, shampoos and, come to think of it, virtually everything in your local supermarket that smells.

Let's get one thing straight right from the outset: these big-boys will not be the least bit interested in the ½ litre of Rosemary oil that you have just produced—even though you may be chuffed to bits with it and keep whipping the cork out to have a sniff. Unless you can produce something unique and spectacularly amazing, the piddling scale of your production will exclude you as a 'potential source of supply'. This means that you have got to go out and sell your own oils or produce your own products and sell them too, either directly or through suitable, hopefully local, outlets. In this way, it's back to the Farmers Market format that is now operating so successfully in Britain. At least you can content yourself with the fact that a major part of the profits are coming directly back to you and aren't simply ending up as shareholders dividends, golden handshakes or part of a Chief Executive's annual bonus.

As for the oils, there are hundreds of the little monsters wafting about and although the majority are unsuited to production in Britain there is still a significant number that can be grown and distilled here. No-one would expect a successful crop of ylang-ylang or geranium or nutmeg to be grown successfully in Britain, but what is wrong with a British Lavender, Clary Sage, Peppermint, Angelica, Caraway or Rosemary, to name just a few?

But would it be financially viable? Possibly. Would it be more financially viable to use the oils to make a range of toiletries? Probably. And would it be even more financially viable if it formed part of a 'package', with maybe a 'Gifte Shoppe' and possibly 'The Olde Rose Garden Tearoom' as well? Personally, I think you would have to really try quite hard to cock that one up . . .

What you **can** do with your precious bottle of oil is to give it some form of attractive packaging and sell it as the **real thing**. Alternatively incorporate into various easy to produce toiletries such as soaps and shampoos or possibly bath preparations, candles etc. or even think up your own novel end use.

Although the correct inclusion rate for these essential oils has to be determined by trial and error, it is usually surprisingly low and will often be less than 0.2%—so you see that your ½ litre of Clary Sage oil is enough for say well over a thousand standard bars of soap. Provided you have suitable outlets for your products and provided they are well presented, then hopefully your sales will soar, thereby justifying the outlay and energy expended. However it's not just your products that should lead to success: it can be the entire concept too (sorry for slipping into that appalling piece of jargon).

Making Sense of Making Scents

Just shut your eyes and imagine . . . it's a warm Sunday in July, the kids are being foul and to make matters worse you're hosting the monthly dutorial visit of Mum and Auntie Rene who are over for the day. They are both in the kitchen trying to be helpful. Somehow they manage to get in the way no matter where they stand and they then insist on bypassing the dishwasher and rinsing their teaspoons and teacups in approximately 40 litres of hot water before drying them up and putting them away.

In both desperation and a last ditch attempt to stay sane, you say brightly "I know! It's such a lovely day: we'll go out for a run after lunch—let's look in the local paper to see what's on. What about Snottlesham Castle?"

"Bor-ing, bor-ing. And it's only ruins. And we got told off last time for climbing on their rotten walls"

"Well what about Sniddleton House and Gardens?"

This time Aunt Rene chips in. "I'm sure we went there two months ago. Don't you remember? The children got into the orchid house and turned all the heating off. The staff were quite nasty about it and said we didn't have them under proper parental control. In fact I think you'll find we've been banned".

As hope fades fast you finally spot it and after whispering a hurried 'Thank you, God', read out: "Here's one we haven't been to—'New! Snippledean Farm Essential Oils. Wander through our fragrant herb and flower fields. Picnic Sites and Children's play area. Come and see our wonderful essential oils being produced in our New Still Room. Tea Rooms and Gift Shop for soaps, bath oils, perfumes . . ."'

I know that this may seem hard to believe, but this urban anguish is being duplicated the length and breadth of Britain on most weekends. The British population, who are now a mainly heathen bunch, quite simply have a job filling up the non-sleeping bit of Sundays. There comes a point, even if you are as slow as a slug, that the DIY and home decoration projects set by the Memsahib finally come to a conclusion, even if this has meant you eventually had to accept defeat and get in a professional. The garden is looking pretty hunky-dory and the thought of yet another visit to those rude Herberts in the Garden Centre is making you rebel.

"No Peggy, I'm not spending another penny in that place until that patronising plonker with the ponytail apologises".

So what do you do? You go out for a ride, that's what. Pretty soon the radius has to be expanded as the list of acceptable places of interest are exhausted, one by one. However, you can't go too far otherwise the wrinkled ones will need several breaks and it will then take over 2½ hours to go 35miles. This is because their leg/back/buttocks are getting stiff, they are too hot/cold/fuzzy-headed or they are suffering from thirst/hunger/wind and need to go to the toilet for a pee/change of underwear/swap cardigan for a different one, take their pill etc. Note that these irritations pale into insignificance compared to what the brats have in store for you: no wonder you have the Samaritans on Speed Dial.

That is why any new attraction—especially a well thought-out one, such as an essential oil distillation plant with attractive gardens, something to keep the kids interested and perhaps a shop and an outlet offering some well presented gifts has such a good chance of succeeding. The still itself is just so *exciting:* it sits there, a gleaming shiny contraption usually 6 or 7 feet tall with hot bits and cold bits, oodles of piping and lots of glass tubes full of oil droplets blobbing about with water trickling around—and it makes the whole place just smell wonderful too: Snippledean Farm's answer to the Traction Engines, only this time with a saleable end-product. Irresistible!

Even if you don't want to take on the whole shebang and completely alter the character of your farm, it is still worth going down the essential oil road and then letting others have the pleasure of marketing and selling your goodies. Put simply, all these items are consumables: I know the Eau de Toilet aftershave given to you by Aunt Rene circa Christmas 1994 may still be hanging around like a bad smell, but smellies normally do get, well, consumed. There are surely enough outlets for the determined soapster to service. Provided that the homework's been done, the market was thoroughly researched, the goods are top notch and the packaging is eye-catching then I'm sure that 'Clarissa's Country Crafts' or any other suitable outlet would be proud to have your products on their shelves.

Finally a brief word about profitability. I would think that one simple example should put things into perspective: let's take a one-acre field planted with lavender. Your yield of oil is around 7kg and you have packed it all in 10ml bottles for retail sale. Your outlay must include the original plant costs plus their care throughout the year, as well as planting and harvesting. Then add on the distillation charges plus cleaning up the oil and the price of decanting the oil into bottles and labelling the product to look attractive. There is also a host of other charges that often get missed including advertising, VAT, origination fees for labelling and design, bank charges, wretched accountant's fees, beer . . .

So what do you get back? Good English lavender oil is currently retailing for around £10 for a 10ml bottle so remove the VAT and you are left with about £8.33. Deduct the cost of the packaging (say 50p) and you give the shop £3 mark up for each bottle sold. Subtract another £1.50 per bottle to cover incidentals and labour costs and you are then left with 7000 x 10ml bottles of Lavender oil @ a paltry £3.33 each. To most farmers today, this still equals lots and lots. Naturally there is a big initial outlay and you've still got to sell the stuff: even so, it should certainly turn out to be as Arthur Daley would say, a "nice little earner" and I would imagine it would give a darn sight better return on that one acre than cows, sheep or pigs ever could.

Chapter 3

Janet and John play at Science

Before we waft away on a wave of fragrant euphoria, let's look at a number of essential oil essentials that all have to be in place before we have lift off and it's all units Go! Go! Go! . . . .

First of all, . . . The Land. How much is required? The answer to that is "It all depends on what sort of enterprise you wish to set up." If you wish to produce masses of oil and have the farm heaving with visitors buying cartloads of smellies then you would need a lot more land than an enterprise that merely distils the oils and produces some soaps and bubble bath for sale locally. In all cases however, the amount of land needed is probably less than you might think. Imagine, for example, a field of say two acres. This would produce a heck of a lot of Lavender—typically a fully mature planting of about 10-12,000 plants, would give a yield of about 15kg of essential oil—and probably more like 100kg of oil if Lavandin is planted instead. Just so that you know, Lavandin (*L. hybida*) is a hybrid of the True Lavender (*Lavendula augustifolia*) and Spike Lavender (*Lavendula latifolia*) and grows particularly well at higher altitudes (Impressed?) Other plants usually produce **much** less but then their value is considerably higher, so it's all swings and roundabouts.

Bruce Dolby

Another factor to bear in mind is how often do you want to run the still, since it seems a bit silly to fork out for all the equipment and then only run the damn thing twice a year. It will probably not be able to operate during the winter months anyway unless you buy in all the necessary raw materials, so realistically it will be working for a few weeks or months of each year as the crops are harvested. Obviously it is always best to stagger the planting so that there is not too much of a bottleneck at any particular time, but if you want your still to be more than an ephemeral showpiece then it pays to have a variety of different crops coming into season over as widely spaced period of time as possible. Remember that whilst some plant raw materials can be stored for distillation at a later date, others will rapidly deteriorate or lose their volatile essential oils, so they should be distilled as soon as they are harvested. Do your homework first. See me.

If I had to put a figure on it (and this really **is** a guestimate) I would imagine that a minimum of ten acres of plants would at least put a still with a 450 litre charge basket through its paces. It would also produce enough essential oil to keep the average soap maker happy for a considerable time and you can then determine more accurately what the demand will be and what plants you should grow. It is worth noting that local farmers may also be happy to cultivate and then supply you with your plant raw materials, especially in these difficult times when it is so hard for them to earn a penny.

Then of course you will need a stillroom—this is what all those outbuildings on your farm have been crying out for. This is detailed in the next chapter, but it doesn't have to be a huge area: you could probably get away with a 10ft x 10ft room, as long as the ceiling was at least 8ft high. It will also need to have good access for the raw materials plus power and water.

Apart from the stillroom, what else is needed for you to produce your star bath-time creations—or at least be able to process and clean up your essential oil? You don't need a laboratory or anything special: just a workshop/shed/room somewhere with a door that can be locked. A worktop is vital, a sink, a few shelves and a cupboard for a few bits of equipment such as some laboratory apparatus (of which more anon), measuring cylinders and a set of scales. An area for your packaging will also be necessary. Other basics include the utilities; heat, water and light, plus perhaps a filing cabinet and somewhere to keep your records and stationery. A fridge would be handy to store your cans of beer as well as your bottles of essential oil and that completes The Production Facility. That's it.

So what are these vital laboratory bits and pieces that will enable you to make the executives at L'Oréal gnash their teeth in fear and trepidation (you wish!)? The first thing that the budding distiller will need is some proper lab. glassware so that he can pretend to be Dr. Bunsen Honeydew: let us start with separating funnels (usually known simply as just 'separators'). Unsurprisingly separators separate: in this case, your precious oil from any water. Preferably you should have a really big one holding 2 or even 5 litres carefully stored away in your cupboard, as well as some smaller ones of capacities between 250ml and 1 litre. (*Figures 14 and 15*) It's preferable to get separators with rubbery plastic taps ('Rotoflo' taps) rather than the ground glass ones, since the latter need greasing to stop them seizing up. Similarly the bung in the top can also be non-glass too.

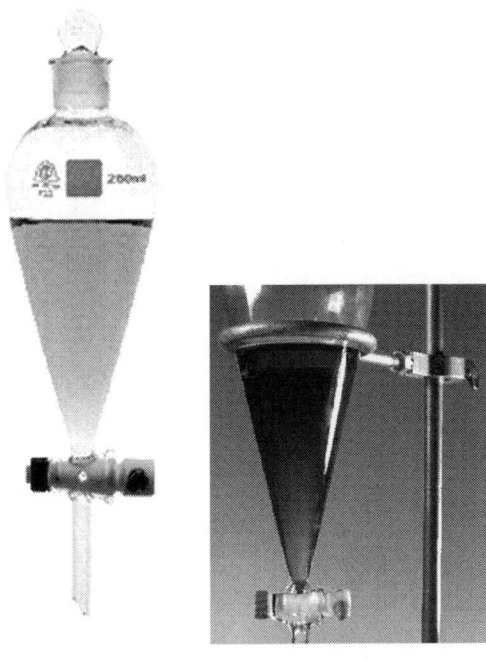

Figure 14   **Figure 15**

A quick word about the sizes here: in any laboratory glassware, the bits that fit together do so via cones and sockets and these are made from the glass and ground to an exact size. Naturally they have to mate pretty accurately, so these cones and sockets come in standard sets of sizes—what a relief, I hear you say . . . Each cone and socket normally has two numbers stamped on them e.g. 24/29—the first number is the maximum diameter and the second number is the length, so if you have a 19/26 cone it will fit into a 19/26 socket and everybody's happy. Naturally there will come a time when different bits don't fit—but there are always adapters available that expand or reduce the socket size so that the cone then does fit the socket. It's easier than Lego.

The separator has to sit in something since it can't balance on its stem like a ballerina. You are going to need a stand. Just like one of your favourite telephone call centres, you now have 4 options:

You can pay silly money and buy a new one from a scientific equipment supplier.

You could get a second-hand one from a second-hand scientific equipment supplier.

Try eBay.

Make one.

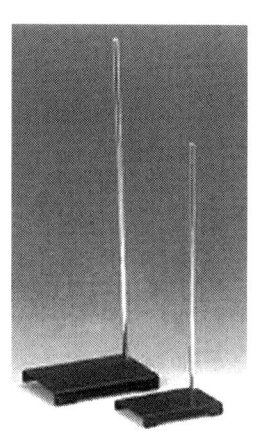

Figure 16

"And of what does this stand comprise, pray?" I hear you ask, to which I reply "Not a lot". It has a base, which can be a flat piece of metal or a flat bit of wood to keep it steady, and an upright, which can be a round piece of metal, or a round piece of wood. For example, a 10mm diameter metal rod, say a metre long would do, or—ah you've guessed—a 10mm diameter wooden dowel, say a metre long. Obviously, if it needs to be on show, it shouldn't look like something your son has brought home from playschool, but it still doesn't need to take a month to make. Typical laboratory ones are shown here (*Figure 16*) and these are particularly posh.

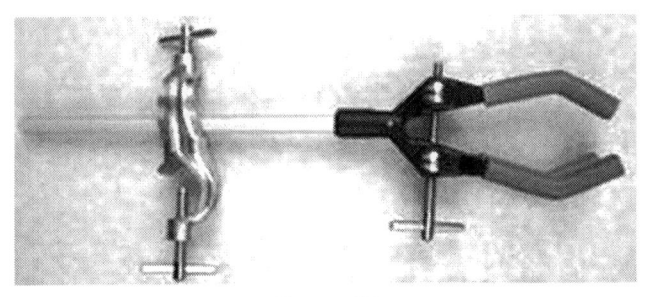

Figure 17

Fitted to the stand is a boss (no, not the Memsahib)—this is simply a device with two threaded bolts that can be done up by hand: one holds it on to the rod and the other is able to hold something else at right angles. I haven't explained that very well, but have a look at the picture and all will be revealed (*Figure 17*)—in this case it is holding a clamp. Although you will need these clamps as well, an O-ring on a rod is better for holding the separator: when this ring is held horizontally the separator will happily sit in it all day long (*Figure 18*).

Making Sense of Making Scents

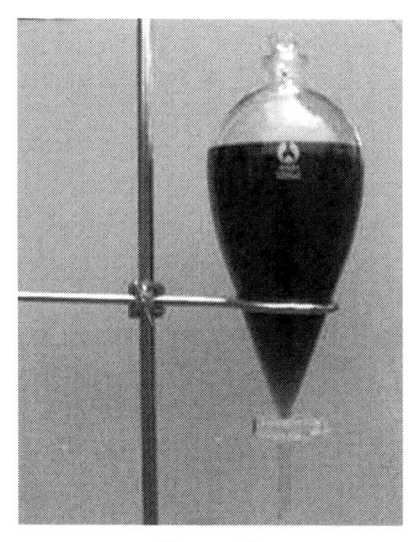

Figure 18

There is an alternative to this system: simply construct a rectangle made from a piece of wood that is wide enough to be stable when it stands on one of its longer sides. On the opposite side cut out some holes so that it resembles a miniature version of one of those multi-seat lavatories much beloved by country folk in Victorian times. You can then sit the separating funnel(s) in their individual loo holes. Sorted.

You will need some anhydrous sodium sulphate crystals to dry your oil (i.e. remove the last vestiges of water from it) and this is added to the separating funnel after the majority of the water has been tapped off. Once dry, the oil is then allowed to pour via the tap into a funnel that contains a filter paper where it goes into a bottle for labelling and storage in your fridge.

Other bits of boffinalia that you will definitely need are standard funnels in a range of sizes, preferably made out of glass and some conical flasks too. It's always best to filter your oil into some nice glass receptacles (*Figure 19*) that do not have 'Property of Bart's Hospital' stamped on them: it also allows you to have a good look at your oil to check that it is free from any water droplets or other crud and is crystal clear. The best flasks are those that can be stoppered tightly to make them airtight.

Figure 19

What other pieces of scientific equipment are needed so that you can operate efficiently? Well, there is one piece of apparatus that I would strongly recommend: it will enable you to determine how much oil is in the sample of plant material so you know what to expect, or if it is even worth steam distilling. It will also show you what the oil in your raw material looks like and what it will smell like. You can even use it to find the best time to harvest your crop to maximise your yield and/or quality. I've always known it as a 'Dean & Stark Apparatus' although this actually relates to

Making Sense of Making Scents

the central glass part and even this bit is now officially called a 'receiver for heavy entrainers' (I much prefer Dean & Stark since it sounds like a nice branch of country solicitors). It is basically a miniature still, comprising four parts. First there is the electric mantle: this is a heating device designed to accommodate a round flask. It looks like a small circular hatbox with a large hemispherical hole in the top and this is where the flask sits. This hole is lined with glorified electric blanket material and there is normally a control switch on the side to adjust the heat and a lead for plugging in to the mains.

Next is the flask: this should be of the alluring round-bottom variety and have a capacity of at least one litre and preferably two—so the heating mantle must be able to cope with this size too. The flask only needs a single socket although if it has extra side one(s) don't worry—simply block them off with appropriately sized stoppers. The main socket should be quite large, however so that you can easily load the flask with your plant material rather than having to poke it down a narrow hole with a biro over a ¾ hour period. With this big socket, you will need an adapter to reduce it down to the size of the cone on the Dean & Stark glassware that then fits on top.

This Dean & Stark apparatus is a simple piece of glassware that has a cone at the bottom and a socket at the top with a v-shaped loop of glass coming off the main glass tube. When the loop is full of water, any essential oil that is present will rise to the surface and collect here, whilst the water goes round the loop and back to the flask. Since this is calibrated you know how much oil you are getting. There are two sizes and the one with a capacity of 3mls should fit the bill nicely: the other size is 12.5mls which is a bit massive and will also be inaccurate to read if you are only expecting less than ½ml of oil.

It may be worth your while getting the other type of Dean & Stark apparatus—i.e. 'a receiver for light entrainers', if you are thinking of producing some essential oils that are heavier than water e.g. clove. This is similar to the other one except that the loop is now a cul de sac and the oil collects at the bottom, where the calibration starts, whilst the condensed water simply goes back from whence it came, i.e. back to the flask.

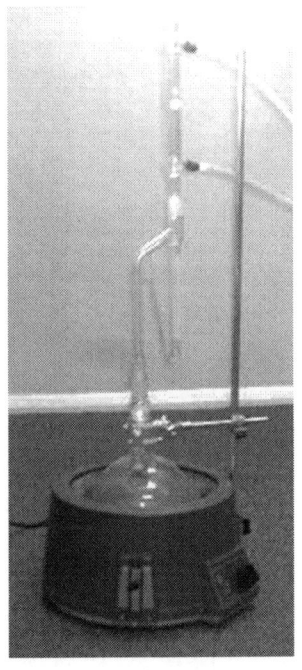

Figure 20

Making Sense of Making Scents

Finally your mini-still requires a condenser. There are many different types available but they all work by the same principle: the flask has been boiled and steam and hopefully a little bit of volatile oil has whizzed up the main tube on the Dean & Stark apparatus. It now reaches the glass condenser in question that is fitted onto the top. This is equipped with an outer jacket; alternatively the condenser could be fitted with a coil or even have both jacket and coil. They are kept cold by tap water running through them via rubber tubes: a return tube takes this water back to the sink but the net result is the same in every case—the steam and oil condenses on the cold surface and drops down into the Dean & Stark apparatus to separate (see *Figure 20*).

That just about sums up all the items that you will need to test your plant materials and clean up your oils. Naturally there will be bits and bobs that I have forgotten but this covers the majority of the laboratory gear that you will need. It can be bought second-hand or on eBay for a tiny fraction of the new price and there are several second-hand equipment dealers who can also assist.

Try and store all your laboratory equipment in drawers and cupboards to keep it clean and out of harms way. For glassware, the drawers should be lined with some form of shock-absorbent material to prevent them from chinking together—the easiest way to get cracks and chips.

If your prize flask does get a crack in it, then don't despair. Provided the damage is not too severe and the flask is not simply smashed to smithereens, it can be repaired by specialist glassblowers who will make it as good as new for a fraction of the cost of the original.

Bruce Dolby

It will also pay you to get some proper laboratory brushes with wire handles, designed specifically for glassware: you can bend them in order to get at the funk lodged in those awkward 'between-the-shoulder-blades' spots that are usually totally unreachable.

Chapter 4

Janet and John make a Smell

Let's cut to the chase and get down to the nitty gritty—let's talk man-talk, let's talk . . . stills. This is the beating heart of your enterprise, the powerhouse, the engine-room, the central and most important part of your very existence . . . sorry, I was getting carried away there. It **is** an important bit anyway.

Historically stills have always been made out of copper, for several very good reasons. Copper is malleable and can be easily beaten into shape, it is relatively cheap and it will put up with lots of abuse during all the heating and bashing. It doesn't react with any of your plants and in fact some people think it may be beneficial in neutralising some of the 'unwanted' acidic or malodorous components in the essential oil . . . but most of all Copper has always been **there**. It has been available; on tap—and even part of tap, as it were. Copper seems to have been around forever. People knew how to work it, they knew how it behaved and they were confident with it. They cooked in it. They bought things with it. They trusted it. How often do we come across Roman coins made of stainless steel or aluminium Elizabethan groats? My case rests.

Today stainless steel does seem to be the material of choice. You might think it bland and boring, but it is inert, strong and looks the part too, especially if you use material with a polished finish (*Figure 21*). The ones we usually manufacture have a 450-500 litre charge basket for the plant material. We started to make stills for the various farmers who decided to have a go at this essential oil craze and this particular size suited their enterprise. There is no reason why we shouldn't be able to produce smaller ones—or larger ones for that matter: the fact is that nobody has bothered to ask us for one.

Figure 21

This is how a still works. Basically, a still is a glorified steamer and is designed so that water is boiled in the bottom of a big vessel, just like boiling a saucepan. It holds a basket with a perforated base that contains your plant material (*Figure 22*) and this is suspended above the boiling water by sitting on a circular ledge about a foot from the base. The steam passes through the perforations in the basket and then through your plant material, hopefully taking some of the volatile essential oils that are present, in much the same way as an elderly relative of mine used to steam Brussels sprouts to feed to any unsuspecting visitors. However, unlike her sprout vapour that used to be released from her saucepan to permeate the house (and next door too), our vapour is funnelled through a huge cone that sits on top of the boiling vessel. This is why the still looks like a miniature chrome plated Oast house—it would double as the perfect present for the child who has everything and is bored with their Wendy house.

Bruce Dolby

Figure 22

From the top of the cone, the vapour is then transported through horizontal stainless steel pipe-work before turning 90º downwards and subsequently joining on to the condenser. Here the vapour has to pass through loads of small tubes (I think there are about 36 of them, but who's counting?), which are all surrounded by cold running water. Naturally the steam plus the essential oil that it carries will condense on these cold surfaces to form a steady trickle, which then gently pours out from the bottom. This liquid is known as the 'condensate' and is basically what all the fuss is about.

The condensate drops directly into a glass contraption that is known for some reason as a Florentine Set: I haven't the faintest idea why it is so named, but it does at least confer a little bit of Italian elegance. The ones that we supply have two 4 inch diameter Pyrex cylinders that are connected to

Making Sense of Making Scents

each other by 10mm glass tubes—the whole lot looks like it would shatter if you so much as think about touching it and that is why we fit it into a sturdy wooden framework (*Figure 23*). The first big cylinder is about a metre long and sits under the condenser to catch the water and oil as it emerges. About an inch up from the bottom of this cylinder is a short 10mm diameter tube welded on to it at right angles—a.k.a. the spigot. Another glass tube is attached to this spigot by a piece of silicon pipe and 'Jubilee' clips—this pipe then goes up vertically for about 85% of the length of the cylinder: and it then bends 90° for a few inches and finally bends down, but only for an inch or so. (Sorry if I'm mixing up metric and imperial—I'm at that awkward age).

Figure 23

What this means is that the cylinder will fill up to the point where the pipe bends over and any surplus will then empty through this tube. The cylinder will always remain about 85% full, which is just as well since this is where most of the essential oil collects, bobbing happily away on the surface like, well, an oil slick. Occasionally one of the little droplets of oil that comes slaloming down the condenser will dive so deeply down the cylinder that it will be able to make a dash for freedom and will whiz down the side tube. It then goes with the flow and plops out from the top of the side tube and into the big wide world. Alas, its freedom is short lived, for here it drops through a short gap . . . and into the second cylinder. Gotcha!

The second cylinder operates in exactly the same manner as the first, and looks similar too, except that it is slightly shorter than its neighbour. It is there to collect any escapees from the first cylinder and also extracts any oil that has been partially solubilised in the hot water but now fancies separating out in this somewhat cooler environment. This cylinder too has a spigot near the base with a glass tube attached to it going vertically upwards to the top of the cylinder but this time it doesn't bend over but has a horizontal side arm instead at a point 85% of the way up and the condensate pours down this arm and into a piece of silicon hose (the top arm of the glass tube is left open, otherwise a siphon would start up and the cylinder would simply empty itself). There is a stainless steel tube welded on the side of the boiling vessel of the still, being joined to it via a 'P' trap, and this is where you poke the other end of your silicon hose (don't ask).

Making Sense of Making Scents

What all this means is that . . . the water goes round and round, whoa-oh oh-oh . . . oh-oh—and it ends up here. 'Here' happens to be back in the still, whilst at the same time a nice juicy layer of oil should be forming at the top of the first cylinder and perhaps a few millimetres in the second one. The water simply circulates ad-infinitum and could even be used again if you are distilling the same type of plants.

The oil is removed from the cylinders by using the glass taps that have been fitted at the front about half way down. The technique is to allow nearly all of the water to come off first and then collect the oil in a separating flask. Next collect the oil from the second cylinder in the same way (if there is any) and you are finally left with your golden fleece—or at least your golden oil.

There is another tap fitted at the very bottom of each cylinder and the purpose of this is to drain off any oil that happens to be heavier than water—quite unusual, but not unknown. The other use of this tap is for the removal of 'Floral Water' or aquasol i.e. the condensate without the oil. Most people would normally think that this water might be useful for watering the tomatoes or perhaps washing down the back yard. Think again! This water is precious, my precious. It is saturated with the oil it was carrying and has a powerful pong in its own right. Surprisingly, it can often smell quite different to the essential oil and can be sold as a bath preparation, air spray or used in other fragrant ways. If the essential oils are a 'nice little earner' then how would you describe the sale of your smelly water? I think the words 'money' and 'old rope' spring to mind.

Figure 24

The stills can be heated in various ways but the most controllable option is to use electric heating elements. It is possible to use other forms of heating e.g. gas burners (*Figure 24)* but this is much slower and less controllable. Another alternative is to have a separate boiler, (possibly in a separate room too) and this in turn provides steam directly, firing it into the base of the still via a perforated tube known as a 'steam sparge'. Anyone who has watched espresso coffee being made will get the gist of how it operates. This system was discounted from our plans because of the huge expense of steam boilers today coupled with all the Health and Safety shenanigans that now accompany them. These include yearly inspections and certifications which I suppose is quite understandable—the thought of a boiler failing is really not something one cares to dwell on. However, all these inspections etc. may prove a tad costly—especially when the farm is sited on a remote island in the Scillies or one of the Outer Hebrides—(they are expensive enough in the very un-remote south east of England).

Making Sense of Making Scents

Because the best place for growing and distilling plants for essential oils is not necessarily the best place for having a 3-phase power supply, we decided to produce stills and use simple single phase 3 kWatt heating elements—the ones that are currently keeping around twenty five million immersion cylinders hot throughout Britain. They should also be perfect for boiling the water in the stills, provided that the thermostats are removed or bypassed. They are also mass-produced and are therefore dirt cheap, costing less than £15 each and should be available from your local ironmongers (if such a shop still exists), or failing that, your nearest D.I.Y. mega-emporium-hypermarket. All that is then needed is to wire them up with the correct heat resistant cable so that each one is able to be individually switched on and off, using a wiring circuit designed to cope with the wattage required—if four elements are used, then about the same amount of electricity and grade of wiring will be required as for a large electric cooker.

Since there **are** normally four heating elements to each still and each one is 3 kWatts, it will require 12 kWatts, although you should be able to switch off two (or even three) once boiling point has been reached. Today of course, you will probably need a fully trained and qualified electrician to even consider putting the plugs on, so don't try this at home, kids. The still will usually have to boil away for an hour or two, depending on the product in the basket, so the cost of running it, whilst not negligible, will only affect the expenses side of the balance sheet by a fraction of one percent. To put it into perspective, a pound or two for the electricity should give oil and products worth hundreds.

Bruce Dolby

Figure 25

If you lag the cone and vessel with some form of heavy insulation, you will be able to speed things up considerably and it will then use less energy to boil it and keep it boiling too (*Figure 25*). On the negative side, it will make your beautiful gleaming still look pants and be a real turn off for any visitors. Perhaps the best solution is to have a nicely lagged overcoat fitted with Velcro fasteners so that you can whip it off when the punters turn up, give them the Full Monty, and then once they've pushed off stick the jacket back on. I **am** talking about the still here.

The still needs water for the actual steam distillation—probably less than 100 litres goes in the pot prior to the basket being fitted. This could be tap water or rainwater but whatever the source, it must definitely be clean, odourless water. I read a book on house building once that stated rather formally "Any water that is to be used for the preparation of mortar should be of potable quality". I don't know what the Brickies thought of that, but if potable water is good enough for pug, then it's certainly good enough for your still too.

However, the real water guzzler is the condenser. This floods through a pipe into the bottom of the condenser and swirls around the inner tubes. On its way to the top, the water has several baffles put in the way to ensure the absolute maximum contact and to optimise the cooling effect, a bit like chicanes in motor racing. The water then comes out of the top spigot . . . and usually goes down the drain. Since this water never comes into contact with the vapour or condensate (which is carefully shielded inside all of those thin-walled stainless steel tubes), the quality of the water surrounding them doesn't matter. It can be rainwater, stream water, in fact distinctly un-potable water for all I care—and as long as it is flowing fast enough to cool down all of the vapour, then it's all right by me.

If you live on a farm in one of those places where water is metered and is an expensive luxury, it is certainly worth getting a holding tank and a chiller unit. This means that the cooling water will circulate too and if all goes to plan, your water usage will then be miniscule. The best place to get one of these is at a farm sale: bulk tanks for milk have chiller units incorporated in them and they should prove to be ideal.

The final bit of kit that you will need is a running block and tackle—unless you have muscles like Arnie Schwarzenegger. A 450+ litre basket will probably hold only 50 kg or so of plant material, depending on its packing density, but that still promises a ricked back or split trousers if you try to lift it into the vessel. It has to slide down 'straight', because the basket is only a slightly smaller diameter than the boiling pot. It also guarantees a torrent of somewhat colourful language when you try to lift it out afterwards with all the plant material sodden with water and twice as heavy. That is why the basket has stainless chains fitted to a lifting ring, or alternatively a central lifting eye so that the basket is ready for the hook from the block and tackle. Then all that is required is a quick heave on the pulley chains—and up she rises. The basket can then be slid to one area where the spent material can be tipped out, ready for carting away and then moved to perhaps a different point for filling with fresh material.

The block and tackle should be positioned directly over the centre of the still, at least a foot higher. On the bend at the top of the still at the exact 'balance point' there is a lifting eye fitted: the cone, top pipe and condenser can be raised as one big lump, pulled to one side out of the way and lowered gently to the ground, so that you can then get at the basket. When this lot goes back onto the boiling vessel, toggle clamps are there, ready to snap shut and hold it all together. Neat.

That, in a nutshell, is the basics of the still. But to see how it will run and run, read on . . . .

Chapter 5

Testing, Testing—Having a Brew

It must have been a mental aberration—one of those senile dimensions they keep talking about: you've actually gone ahead and bought a still. Must have been mad. Now you have got it, what are you supposed to do with it? It's sitting in the shed like an obsolete space capsule, the electrician has done his bit, accompanied by plenty of muttering and head scratching. The plumber has sorted out the cooling water and fixed all the taps and drainage. In preparation, some of the old barley fields were planted up last year with rows and rows of Rosemary, plus some Sage and Lavender and an area growing Lord-Knows-What (if Michael had anything to do with it, then it's probably cannabis). Already Shirley is coming out with damn-fool suggestions about tables and seating and where she thinks you should put the crockery and should the shelves for presentation go over there, behind the glass display cases . . . .

Whoa, hold on everyone! Let's take a step back and see what you should be doing now and then try to get some form of system up and running. Naturally, with time you will find your own way to run things, but it will certainly pay to think it out first so that you have some sort of plan to work to.

Obviously the crop has got to be harvested and this is likely to involve a fair amount of graft, on a par with chain gang rock breaking, especially at the start when everything is likely to be done by hand. But first, we need to know if there is any oil actually present and ready for distilling.

So let's go into the Quality Control Laboratory (formerly known as the Fertilizer Shed) where we can fall back on our dear old Dean & Stark apparatus (no, not literally), carefully preserved from Chapter 3. When the plants are well grown and appear to be pretty fragrant but are not yet quite fully 'out', then it's time for a brew, i.e. time to take a random sample and bring it in for testing. Chop up the leaves or flowers (or whatever bit you are distilling) and weigh out a realistic amount—for example, approximately 100 grams—into a preferably 2 litre sized round-bottomed flask. I used to suggest a cheap option for accurate scales being the type designed for weighing postage items in offices: they are accurate to either 1 or 2 grams, which is certainly good enough for this job. However, today there seems to be loads of unbelievably cheap accurate digital scales available: the demand for such items may possibly come from all those gentlemen who sell their wares on street corners. Every cloud has a silver lining . . . .

Sit the flask in a support ring on the weighing pan, tare the scales and then add your plant material up to the required weight (or until you get bored).

Once your sample is in the flask, add tap water until it is roughly half full, plonk the flask in the mantle and place it on the base of the laboratory stand. The adapter (if required) is then fitted and the Dean& Stark apparatus placed on top.

Finally your condenser fits on top of the Dean & Stark, remembering that a little silicon grease must be smeared on all glass sockets to prevent the joints seizing together. There is of course proper laboratory silicon grease designed specifically for this purpose . . . or you can also get away with the silicon grease that plumbers use for fitting waste pipes. Finally, at a pinch you can even use Vaseline, or any of the other lubey products that you find at Boots.

As this set up is a bit top-heavy, it will need clamping, preferably twice: once around the neck of the flask and again half way up the condenser. Here you use the clamps in the bosses, remembering not to do them up so tightly that the glassware smashes like a walnut in a nutcracker. A gentle girly grip is all that is needed: yes Michael, **now** is the time to get in touch with your feminine side.

Next the rubber tube is then affixed to the condenser for the cooling water: there are two ways in which this is done. In some of the newer condensers there are two little separate glass spigots for the rubber pipe and these also have plastic screw fittings. These fix them to the condenser, which fortunately also has threaded inlets/outlets to accept them. In the second (original) type the rubber tube fits straight onto the condenser's inlet and outlet point. I presume we have Health & Safety to thank for that: in the latter case, there is the faintest possibility that the same Neanderthal who over-tightened the clamp and smashed your flask could also try and force a narrow rubber tube over the glass spigot without bothering to lubricate it, or warm the rubber pipe first. Result—possible broken glass, stigmata and subsequent claims for compensation. In the first system, this is not possible; although thinking about it, a child could possibly eat these small glass/plastic attachments. I sincerely hope that sufficient warnings to cover such an eventuality are now included with the packaging of each purchase.

The rubber tube from the bottom of the condenser fits on to the cold-water tap whilst the top tube goes back to the sink. Turn on the tap so that a trickle comes out of the return pipe, pour a little water down the condenser to fill the Dean & Stark apparatus, so as to catch the very first oily droplets—and we are ready to rock. Turn up the mantle heat to maximum and wait for it to boil. As it nears boiling point, turn down the heat to achieve a gentle wobbly simmer . . . and then just watch.

And Lo! There is oil! Texas Tea. The Very Chrism. Okay, okay,—but at least it shows it is **there**.

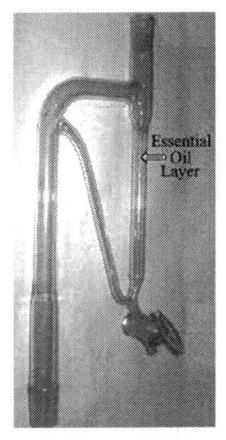

Figure 26

Dripping off the condenser will be a steady drizzle of condensate, merrily splashing down into the Dean &

Making Sense of Making Scents

Stark—and there, sitting at the top is a layer of oil slowly forming. It may be boringly colourless, but depending on the plant material it could also be one of many colours; often it is golden yellow, or pale green or even a delicate faint blue (*Figure 26*). Exciting, huh? The layer is hopefully getting bigger too: depending upon the oil content of the plant material you are working on, you could end up with anything up to 1ml or more, or even 2 or 3 mls plus in the case of a few plants such as Lavandin. However, 0.2 to 0.5 ml is much more likely so don't be too disappointed if your harvest turns out to be just a micro-squirt. The clever little Dean & Stark apparatus has a set of gradation marks down the side and enables you to determine how much oil you are getting and when you have got out all that there is—the layer simply stops growing. This could be 20 minutes or as long as two hours—simply take a note of how much oil you have got and then 10 minutes later see if there is any more. If the layer has stopped growing, then I'm afraid that's your lot, chum.

Ain't Science Grand!

So if you have put 88 grams of Rosemary flowers in the flask and you have ended up with 0.35 ml of oil, then your yield is a respectable 0.40%. Oh, for goodness sake!—it's hardly Honours Degree level Mathematics—what **did** you learn at school? Sorry, I didn't know that you never went. Anyway, simply divide the amount of oil by the weight of plant material and multiply by 100 to get your percentage.

If you try this test a few days later it will hopefully show that your plants now contain a bit more oil. Perhaps they have already reached their optimum level so then you should harvest without delay. With time, you will soon discover the best time to distil your plants and this could well be by simply checking their physical appearance, or by sniffing them, or tasting them, or smoking them . . . you will eventually have that indefinable countryman's (or woman's) fey nous and you'll just **know** when they are ready. Nobody will question . . . . Big John. When he says distil, we distil.

I would like to pass on a few tips to assist in the teething days of your Boffinhood:—

If the condenser feels warm or hot during the operation, slightly increase the flow of cooling water; otherwise you may lose some of the oil as vapour out the top. Actually this can be a great idea to improve the aroma of your room and hide the smell of stale beer and those trainers, (although it's a terrible idea for the accuracy of your oil yield).

Occasionally you may find that the flask does not boil smoothly, but lies dormant for a few seconds and then 'bumps'. This is caused by the smooth internal surface of the flask not providing enough 'rough edges' for the steam bubbles to form on, so superheating then occurs. Adding a few bits of something porous at the start e.g. broken flowerpot, will prevent it happening. More scientifically, anti-bumping granules are there for the job (yes, they actually exist).

If you happen to be growing a plant that has an essential oil heavier than water, then the other type of Dean & Stark apparatus will need to be used, although the operating system is identical. This time the oil collects at the bottom of the arm in a glass cul-de-sac and again there are gradation marks to see how much you have collected.

Very occasionally you get a pig of an oil that can't make up its mind if it's heavier or lighter than water—some collects on the top whilst a few globules decide to go down the tube and back into the flask. Here you have two options—firstly you can try and keep the water in the Dean & Stark very cold by placing it in a jug full of ice water whilst in operation and hope that the lower temperature will change the specific gravity in your favour. With luck the oil will then all stay floating. Alternatively you can add a known amount of a very light non-miscible solvent e.g. Hexane, so that it sits on the surface of the water in the Dean & Stark apparatus and collects all the oil as it passes through. Deduct this volume added when calculating your yield.

The oil can be removed from the Dean & Stark apparatus with a little instrument known as a dropper pipette: this comprises a thin tube drawn out at one end into an even thinner proboscis and fitted with a rubber bulb on the other end. There are even plastic disposable ones costing a few pence each that are made all in one piece and these are calibrated too *(Figure 27)*. Simply squeeze the bulb, stick the narrow end in the oil and let it suck it up by releasing the pressure. There are even some Dean & Stark kits around with a glass tap at the bottom so you can drain your oil out at the end. Clever.

Bruce Dolby

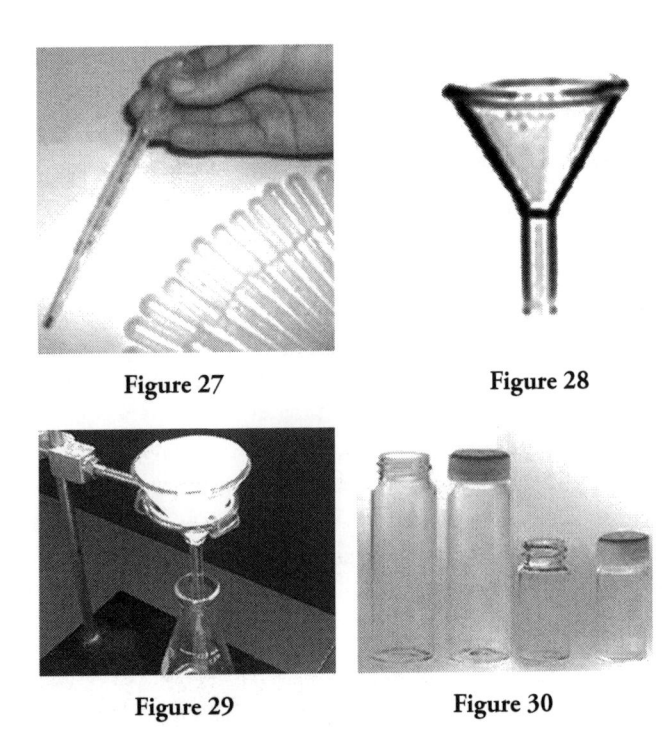

Figure 27

Figure 28

Figure 29

Figure 30

You may want to check out this oil for fragrance, although it may not be quite the same or as good as the oil you get from the big still, which only passes steam through the plant material. Here we are giving it a prolonged boiling in water in the style that was perfected with cabbage by cooks in school canteens. Place the oil into a small glass vial and add a pinch of anhydrous sodium sulphate to remove any water and then pour it through a tiny funnel (*Figure 28*) lined with an even tinier filter paper (*Figure 29*) and collect it in another vial (*Figure 30*). Once you have got your dry oil sample, you can now inhale as deeply as you like (unlike Bill Clinton).

This small-scale trial is identical to what will happen when you start distilling in your main still—hopefully you should be working on a larger scale, but the same principles apply.

It is vitally important that you keep records—and we are not talking about your set of Donovan 45's that attracted zero bids on eBay (even without a reserve). At least keep a daily log of what you did, how much oil you got, what problems you experienced, the address of a supplier you discovered—in fact please record **everything**. The book (or computer file for the more adventurous) will be your Bible and until you are totally au fait with all aspects of essential oil production you will refer to it, add to it and rely on it for years to come. Much later this record will provide you with a wonderful reminder, covering all of your successes right across the spectrum, even down to distillations that come under the category 'EMBARRASSING. DO NOT REPEAT. EVER.'

Similarly you should keep samples of your oil for reference too, preferably out of strong sunlight and kept nice and cool. It's best to label them with a code so you can look up all the relevant info in your book/file. There may be simply too much information to fit on a tiny label anyway and nothing puts off the punters more than a glass vial on display with the comment "Smells like Cat's Pee" emblazoned on it. This is also where your storage capacity and filing system comes into its own, because if you are not careful, you will end up with a logistical nightmare. It's a pity that fragrances can't be downloaded straight into a computer too, but until then you'll just have to store dozens of samples in little glass vials.

Sorry, never mind. Live with it.

Chapter 6

Doing it for Real

The time has come to put your still through its paces and for it to start earning its keep. After all, you have been sitting there straining at the leash for weeks whilst the still has just sat there, pretending to be a NASA centrepiece. You have spent months with all the cultivation, rebuilding, plumbing, electrics and Lord-knows-what, spending a small fortune in the process and not one Euro-cent in return. It's payback time.

Your crop is ready, it has been tested and the essential oil is all present and correct, Sir. So let's be 'aving it!

Figure 31

At some point you will have to consider a mechanised method of harvesting your crop. It needn't be an arm and a leg job either: here is a simple but extremely efficient lavender harvester, driven by two doughty chaps from Carshalton Lavender. I know it looks like it is powered by a Strimmer engine, but it sure as hell beats cutting by hand . . . *(Figure 31)*.

Presumably your crop will be brought to the stillroom by a tractor and trailer and this is where you reap all the rewards from working out the correct design first.

The trailer has been backed up to an area adjacent and above the point where the charge basket is sitting. The plant material is slid out the back of the trailer and on

Making Sense of Making Scents

to a clean platform area that has been built in the correct location, out of the rain. The charge basket has now been lined with muslin with a reasonable length of about one metre over-hanging the edges. This is to stop small bits of your plant material e.g. Lavender or Rosemary or whatever from dropping into the water and contaminating it: unlike with the Dean & Stark apparatus, we only want steam to go through our plants: we certainly don't want to boil the living daylights out of them.

The next operation is to fill the charge basket with your plant material. It should now be easy to just slide it off the deck and straight into the tub. At last all that work is hopefully coming to a successful finale and expectations are rising. Your very first distillation: hearts are a-beating and fingers are crossed. The atmosphere is electric. I'm getting quite excited just thinking about it and I don't even know you!

It is worth packing it properly so don't just shove it all in and stamp it down—you're not in a French vineyard. You want the steam to pass evenly through the whole tubful so you should try to aim for a nice even packing density right up to the top. It shouldn't be filled so tightly that the steam has a job getting through, but you still want to maximise the charge that you can get in. The ideal is to find the optimum amount for a decent yield and at the same time get a speedy distillation. This is where you will have to do your own trials to perfect the process.

Once the basket is full, fold the surplus muslin over the top and tuck it in. What you have basically produced is an enormous haggis that your Scottish grandmother would approve of. The top surface should now be reasonably flat so that the steam will whiz through and take the essential oil with it. It's ready for lift off.

You will need some scales so that you can weigh the basket before and after packing to find out how much plant weight you have put inside: that way you can determine the yield of oil.

Every week in the Angling Times there is a constant array of photos of blokes in bobble hats with cheesy grins, arms outstretched as they cuddle 82lbs 5ozs of slime—a.k.a. Tope/Carp/Pike/Bert. They know it is exactly 82lbs 5ozs because they have a spring balance with a whacking great hook on the end. Get one. If they are perfect for weighing a shark then they will do nicely for 53kg of Clary Sage, which hopefully shouldn't be thrashing about trying to exact revenge.

After the weigh-in, slide the shoe of the block and tackle over to the basket and pull on the chain to lower the hook. Carefully lift the basket up and then slide it along to the open vessel. This has been sited directly under the block and tackle and filled with tap water so that the electric elements are covered by about two or three inches. Lower the basket gently into the still vessel until it sits tight on the internal shelf. It may pay to have a helper here nudging and twisting the basket if necessary, since it is a pretty tight fit. The hook is then removed from the steam pudding and slid (hopefully in the opposite direction) to where the cone-plus-top-pipe-plus-condenser ensemble is sited. It's a good idea to make up a simple wooden frame on which this bit of kit will sit: it stops it getting damaged or dirty and makes it

Making Sense of Making Scents

easy to clean too. Fix the hook into the lifting eye at the top and haul away on the chain. It should rise up smoothly and after much see-sawing come to rest horizontally. Here you **will** need a helper. Slide this contraption over to the still vessel and lower it so that the flat sealed surface of the vessel mates with the flange on the cone. If all is in the correct orientation the six stainless toggle clips will fit together perfectly and will be able to be snapped shut. Both sections are now fully secured into one complete distillation unit.

It is probably a good idea to give extra support to the still by leaving the hook in place, fixed into the upper lifting eye: alternatively a support can be made that takes the weight at the condenser end. However, since the whole lot now weighs probably over 200 kg it is unlikely to be going anywhere—but it still looks a bit lop-sided so it's reassuring to have the extra security.

If the two lengths of rubber water pipes joining the condenser to the tap or going back to the drain are long enough then they can be left joined on whilst everything is moved into position. If they are not, you now have two scenarios. (1) You can undo the jubilee clips and pull the pipes off each time the condenser plus cone unit is moved and then have to put them back on again ready for the next distillation. (2) What will actually happen: one day you will forget that the pipes are still joined on, so that as you slide the condenser ensemble to one side it stretches the rubber pipes to breaking point à la Tom & Jerry cartoons. One of the rubber pipes will then snap with a satisfying twang, pouring water in all directions. It will then either smash into your delicate glassware and wreak havoc, or it will 'Stripe you round the mooey, Gord save you if it don't' to use the colourful vernacular of some travelling folk I once knew. Personally I think it best to have long enough lengths of rubber hose.

The Florentine glassware in its wooden frame is now put in place so that the end of the condenser is positioned directly above the larger of the two cylinders, usually with a funnel fitted in between. The Florentine frame should be placed as high as possible without the glassware or funnel actually touching the condenser so it will probably need to sit on a small platform or set of bricks that have been fixed in position first. If the Florentine frame can also sit in a recess that has been designed to hold it, then it will be even more securely positioned with far less risk of any accidents occurring.

The Florentine's flexible exit pipe should then be routed back towards the still. There is a 1½inch diameter stainless steel tube that comes out from the base of the still and then goes vertically up the side of the vessel—this is where your pipe is headed. It is best to support the exit pipe so that it is held in a straight line, with a slight fall downhill right to the point where it disappears into the stainless steel tube. If necessary saw a bit off the tube to shorten it so that this gentle fall is achieved. Remember that the pipe will be taking reasonably hot water back to the still and will therefore become very floppy—hence its need for support (and possibly counselling too). It only needs to poke down the stainless tube for a few inches.

Right. That's it! Let it roll. Fill up the larger Florentine cylinders with water and note that the surplus will cascade through its side-pipe and into the second cylinder—and when that is full, down the exit pipe and into the still. That's the first bit done. "Florentine ready to accept oil, Sir". Now turn on the cold water to the condenser so that a trickle appears from the return pipe. "Coolant flowing. Condenser at low temperature, Sir" Finally turn on all the heaters. "Maximum heating applied, Sir. We have lift-off". Now, if the electrician has done his job properly . . . nothing should happen. Well, what did you expect? It's hardly going to take off and blast into orbit.

However, a few minutes later, when you cautiously feel the bottom of the still, it is definitely hot—in fact the whole area has that 'new hot metal' smell and soon the still vessel is too hot to touch. Now there is the sound of something approaching boiling point; the same noise your kettle makes, only louder. We're getting there! Suddenly you notice a faint aromatic whiff. Everyone peers at the end of the condenser and suddenly a few drops of water appear, splashing into the Florentine cylinder. Yes, these are your first few drops of condensate and soon there is another cascade, followed by another and another, at shorter and shorter intervals until condensate is actually trickling down like a tap just turned on. Don't be so surprised that it works—that's what it was supposed to do!

Time to turn off one of the heating elements and possibly even two: provided it still boils away nicely then use the minimum power necessary—why waste electricity? But just look at the Florentine! There is already about two centimetres of oil floating on the surface and the stillroom is now filled with a fabulous aroma too. For goodness sake take that silly grin off your face (although after all that effort you put in, it must be hard not to feel elated).

By now the oil is cascading down into the cylinder in golden droplets and rising quickly to the top: there must be a couple of inches there already. You had better go and get Michael so that he can see the results of all his cultivation and tell Shirley the still is running and at last there is something to see!

The water is happily pouring out of the first cylinder through the glass tube that comes out the bottom and flowing into the second cylinder—there is even a tiny layer of oil formed here as if by magic, because you haven't seen any oily droplets going up the tube. (The oil is actually separating out of solution because this cylinder is at a lower temperature and is then rising to the surface). The water is also flowing out of the exit pipe from the second cylinder and disappearing back into the still so at least you don't have to worry about the still boiling dry.

Once equilibrium is achieved, you can quite happily leave the still to run without any intervention. Over the last hour and a half, the oil layer has been slowly growing and is several inches deep now. Mark the cylinder at the base of the oil layer with a marker pen and go and have a cup of tea or coffee, or for this inaugural distillation, you may want to stretch to a celebratory glass of Lambrusco—after all, you've earned it! Twenty minutes later when you check it's still happily boiling away with the condensate trickling round and round but the level of oil has not increased by more than a millimetre, it's time to call it a day. Switch off the remaining heating elements and when the condensate stops flowing, turn off the cooling water too.

So what have you got? Apart from several litres of 'Floral Water' there is a reasonable quantity of pure essential oil that could be worth quite a lot of money, either as it stands or as a fragrance for your products. In addition, if you are using the still as part of your 'Essential Oil Experience' then you have a top of the bill showpiece that should keep the visitors entranced for hours.

When you actually think about it, this distilling lark is easy peasy: you just stick in the plants and turn on the electricity and water. There are however a few points that you should at least be aware of. Firstly check the temperature of the condenser regularly. You don't have to stick a thermometer up one of its many orifices; simply feel the outside jacket. It should always be really cold and if it's not then the flow of cooling water will need to be increased. Otherwise you could lose valuable oil as vapour and accidentally turn your shed into an aromatherapy salon. You may also notice a smoky mist appearing at the condenser cone at the very start of the distilling process and this vanishes as soon as the condensate starts flowing properly. What you are seeing is a tiny quantity of the most volatile components in the plants' essential oil coming off first. Heady stuff indeed, but unless you have access to equipment that will cool the condenser down to a very low temperature e.g. -20ºC, then you will have to forget about these few drops and give them their freedom. In the scheme of things it is certainly not worth worrying about.

Finally a few words of warning. What you have here is a pretty hefty lump of equipment, a large part of it being kept at boiling point. There is also a complex set of glass cylinders and tubes that cost several hundred pounds and these don't take too kindly to either dogs or footballs. Rubber tubing may be kept out of the way by being hooked onto supports etc., but no matter how hard you try there will always be an occasion when some nerd will attempt to trip up on it. There is also a drainage tap at the bottom of the vessel and if small mischievous fingers do manage to turn it on, it will rapidly let out all the (possibly boiling) water, all over the floor. If the still is operating at the time, the elements will then become exposed and will probably burn out too: that is why the drainage tap has a facility for it to be padlocked shut.

Never has a securely roped off visitor area been more appropriate. Do use common sense in the stillroom and get your local Environmental Health or Health & Safety people involved too. It is probable that warning notices will have to be put up in various working sectors whilst other areas must be kept securely visitor free, as well as free from fiddling fingers. It will certainly pay to keep all authorities on side—and definitely cheaper than the alternative. However, the sight of a still in action and actually producing essential oil will be a truly fascinating and fragrant highlight for many visitors and the stillroom can really act as a focal point for the whole farm. It is also rather satisfying to explain to interested people exactly how everything works and then letting them smell the final product.

Of course there **has** to be one big negative. How to cope with the clever Dick in a baseball cap who thinks he's the first person ever to shout out "Oi, mate! Where's the whisky come out then?" and who then decks up at his own sublime wit. You will have to have your own put-down ready for this inevitable occasion. Failing that keep a can of Mace handy.

Chapter 7

Now What? . . .

There's Oil in them thar Stills. And to prove it, there it is; it's floating serenely at the top of your Florentine cylinder(s) like, well, oil I suppose.

As those of you with teenage offspring will know only too well, the next comment has to be . . . "**So?**"

So . . . at the end of a fiendishly long slog, you have at last got some precious oil: now is the time to separate it from its watery accompaniment and when at last it is pure and bright, to capture it, seal it in a bottle and then, like Aladdin's genie, hide it away until ready for its moment of release. All right, perhaps I did get a bit Catherine Cookson there, but many essentials oils **do** keep their colour better if stored in dark glass bottles, or at least out of direct sunlight. In any case, essential oils should always be stored in glass since many eat plastics for breakfast, given half a chance. Anyway, we have got quite a way to go before we reach that stage, so let's go back to the basic procedure of how to clean up and bottle your oil.

It is currently floating on the aqueous layer of condensate (or 'hydrosol' to give it its new fashionable handle) in the first cylinder of your Florentine kit: there may be a little bit in the second cylinder and this must be collected too. The still has been turned off for half an hour and the cylinders are now virtually cold: time for a harvesting, time for a quick reap. Carefully hold a suitable glass container under your cylinder's central tap and gently drain off the aqueous layer—sorry, hydrosol. As the level of the oil approaches the tap hole, slow the flow down to a minute trickle and finally stop it as the oil reaches the point just before it would come out. Repeat this procedure for the second cylinder and then carefully stopper your hydrosol container; as you know, this aqueous material will have a significant value in its own right.

Now for the hard stuff. Take your glass separator and hold its socket under the cylinder tap and let the essential oil drain in. Golden rule number 1: Always double and even triple check that the bottom tap on the separator is turned off! It's both heartbreaking and incredibly smelly to have the fruits of your hard earned labour cascade over your crutch, just because a stupid piece of glass or plastic is facing the wrong way.

Assuming that you have successfully negotiated that hurdle, repeat with the oil in the second cylinder if there is any, adding it to the first lot. Place the separator in the support ring fitted on your laboratory stand and have a good look. Hopefully you can now see a fair quantity of oil sitting there—perhaps a couple of litres or more, with just a little bit of water at the bottom that has accompanied it. Carefully drain off this aqueous layer into a beaker and add it to the rest of the hydrosol that is sitting in its glass container.

For those few oils that are heavier than water, these will collect at the bottom of your Florentine cylinder. There is a tap there for just this eventuality—this time the oil will come off first, followed by the aqueous layer and you will have to pour it from one separator to another in order to remove the last of the hydrosol.

Your oil is now visibly free from water, but it will be still saturated and will therefore need dehydrating. If you don't dry it, your oil can possibly hydrolyse with time i.e. it may react chemically with the water it contains. This hydrolysation process can, with time, create some unusual off-notes in your oil that range from the smell of death to the merely sulphurous hum encountered after a night of Mackeson and pickled eggs. In all circumstances, it's definitely best to dry the oil.

Golden rule number 2: always keep a bung in the top of your separator, **except** when you let liquid out from the bottom tap: the stopper should then be raised a little to let the air in. Keeping the separator sealed at other times will stop any dirt or insects dropping in although the latter will normally steer well clear, since many essential oils are potent insecticides. It will also prevent your essential oil from losing any of the most volatile components through evaporation. If you forget and leave the stopper in place whilst you are draining, as the water tries to pour out the air will simultaneously try to rush in, causing a massive 'bloop' and mix everything back together again. Expletive deleted.

Now place a funnel (preferably with a wide stem) in the top of the separator and shovel in about two teaspoonsful of anhydrous sodium sulphate, a simple salt that is quite harmless. It is a white powder that is drier than a Stornoway Sunday School and will remove the last vestiges of water from your oil whilst itself becoming the hydrated salt in the process. Replace the stopper and gently shake and swirl the contents for a few seconds; then, with the stopper firmly in place, hold the separator upside down and open the tap. There will be a small rush of air as the pressure equilibrates—and then immediately shut the tap and turn the separator the right way up. Repeat this procedure a few times more until no further air escapes when the tap is opened. By now some of the sodium sulphate will have absorbed the water and taken on a 'crystalline' appearance of its hydrated form, whilst your oil should have become absolutely clear and hopefully as bright as a detergent advert. If the oil is still slightly murky then give the sodium sulphate a little more time to do its job and then repeat the swirling etc. Finally if all else fails, add a little more of the desiccant—that should do the trick.

Next you must get rid of the sodium sulphate—you've only just put it in but now is the time to get rid of it! Obviously you are going to need a funnel and you will also need a filter paper and a sealeable flask in which to store the finished product. There is a knack to folding the filter paper and it is done for a valid reason and not just to show off or prove that you once worked as a waiter at the Mumtaz Indian restaurant and still know how to origami-ise a serviette.

Use a circular filter-paper disc with a diameter large enough for it to roughly come up to the top of the funnel if it is folded into a cone shape (i.e. fold in half into a semicircle, then into quarters and opened out). That is the simple way to fold a filter, but you will often find that the essential oil will only have the very bottom bit of the filter paper over the spout to actually escape through; the rest of the paper has stuck to the glass and is not actually filtering anything. Net result: your oil finally finishes filtering just as you finish the bottle of Ouzo bought the holiday before last and 'Sailing By' comes on the radio prior to the late night shipping forecast. Definitely semi-pro.

This is a better way—fold a virgin disc in half, unfold it, turn it 90º and fold in half again in the same direction. Repeat twice more folding in the same direction but between the quarters so that you end up with your disc with 8 creased segments. Now the classy bit: fold by hand in the opposite direction in between these creases forming a fluted cone shape (*Figure 32*). This has a vastly improved surface area and is therefore capable of very much more rapid filtration. More importantly it has the advantage of giving your lab-cred a massive boost.

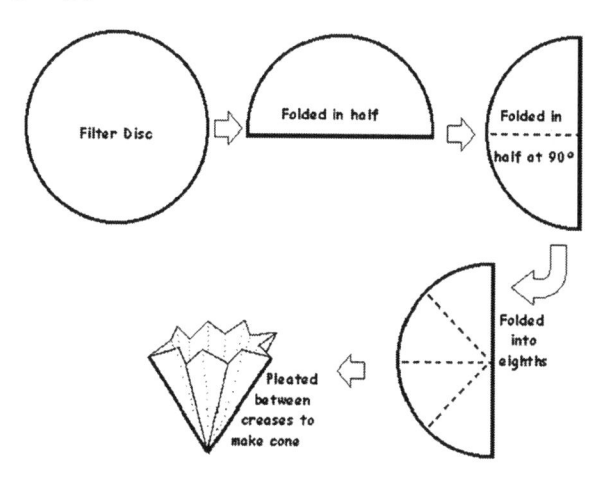

Figure 32

Turn the tap to allow the oil to pour into the centre of this fluted filter and allow it to collect in a suitable glass receiver. Be careful not to let it flow in too fast or it will overflow the filter paper and then unfiltered oil will then mix with your pure stuff. When the last drops have been collected, fit the bung and apply the label, giving its type, date and any other relevant information. Store it safely in a nice cool area and then fill all the details in your record book. At last you can take that worried look off your face.

If it's an expensive oil, it may pay to rinse the separator, funnel and paper with a suitable solvent or diluent that you are going to use later in one of your products and then save this fragrant brew for later use.

What about your beautiful big bottle of carefully collected condensate? The famous hydrosol cannot just be left to sit there—it needs to be processed in its own right too, otherwise it will soon become a less than fragrant broth suitable for germ warfare. Yes, it may have originated from steam, but it is now an ideal growth medium for all the unpleasant little microbial bugs floating invisibly throughout your stillroom and they are parachuting in, ready for rampant multiplication even as you read this. They must be stopped! First, filter this watery tackle in exactly the same way that you filtered your essential oil; through a funnel holding a new fluted filter paper and letting it pour into a clean glass container.

Once this is finished we will need to add a little something to stop the rot. There are many chemicals around that will preserve your solution indefinitely and one of them should be added without delay. They have trade names like Nipasol

Making Sense of Making Scents

or Bronopol and each one has its own inoculation rate. First work out how much hydrosol you have actually got and then pour out the appropriate quantity of preservative into a measuring cylinder according to the supplier's recommendations. Normally the inclusion rates are very low e.g. 0.2% or less and these preservatives simply need to be added and thoroughly mixed. Fix the stopper to make sure it is airtight and label it up. Finally remove beer(s) from fridge and put your feet up. Sorted. Job done.

Sorry, but I'm afraid it's not quite done. A bit like the aftermath of a dinner party, after all the glamour comes the hoovering and washing up. Put that can of supermarket economy lager down: the still needs emptying first. Now that it is cold enough, undo the clamps, carefully raise the cone and condenser section using the block and tackle and slide it to one side: hopefully it will refit on the stand that you have made for it. Next slide the hook back to the centre and lower it so that you can lift up the basket. Pull back the muslin, raise the basket and pull to one side. Now, if you have got your design right, you should be able to empty the contents of the basket directly on to for example a large 'tray' on wheels or a tarpaulin—in fact anything that enables you to easily get shot of this exhausted material out of the stillroom.

Unfortunately the only use that I know of for all this crud is for making excellent compost—which is just as well, since you will soon have a mountain of it. On the plus side, having been steamed it should already be well on the way to being broken down and should be ready in a few months for ploughing back in.

The glassware will also need carefully washing: there are special detergents for cleaning laboratory equipment but washing up liquid and hot water usually works. The taps on the separator and on the Florentine cylinders should be removed, cleaned, re-greased and then reassembled. The charge basket should be hosed down, as should the internal surfaces of the cone. The muslin liner will probably need sticking in the washing machine. **Now** you can have that beer.

So what do you do with your first 5kg of home brewed Lavender oil that is now proudly glistening brightly in a glass demijohn? First of all, you look after it. Never allow a large spanner for example to accidentally drop on it, since the result is a **very** expensive patch of floor which will have an overpowering and penetrating aroma with the capacity to be handed down through several generations. The culprit, who by now will have been nominated as 'The Clearer-Upper' will then need several baths of the total immersion type favoured by religious sects who go in for extreme-baptisms. His aroma will by then have reduced to the merely eye-watering and a prolonged period of time will elapse before his colleagues/spouse/offspring cease crossing the road when they see him coming.

Or so I am told . . .

Chapter 8

What Now? ...

At one of my old work places each perfumer used to be surrounded by zillions of bottles, pots of essential oils and other perfumery ingredients, arranged in rows in a huge semi-circle like an enormous Wurlitzer—which is presumably why they called them their 'Organ'. This was their inner sanctum where they would sit in the centre, like a spider in its web, occasionally taking small amounts for blending. Then perhaps finding a bottle almost at random, dunking a smelling strip before taking a snort and theatrically tossing the strip into the bin.

Presumably it must have been compulsory for perfumers to wear avant-garde clothes in those days in order that they could be creative and be seen to be creative. I used to be totally in awe of them—me in my grubby lab coat whilst they wafted about in a cloud of creativity, dressed in flowery shirts, bow ties or possibly velvet trousers. It's only with hindsight that I realise they could collectively have been summed up as 'devotees of self abuse'.

I was once asked to take a new product that I had synthesised over to the Dutch perfumer-in-chief for his comments. "Broooze—I like ziss ver mudge. It schmells of Mel-lon peeps . . ." There was me, blindly wandering through life in total ignorance, never even knew what melon pips smelt like until then—or even that they had their own distinctive whiff. I soon realised that the world of creative perfumery would forever remain an unfathomable mystery and I was best out of it. However, the simple mixing of ingredients to make perfumed products should be within anyone's capability (even a Philistine like me).

To start with I hardly expect a novice to create a complete 'organ' of essential oils and related products . . . perhaps an aromatic Stylophone or Harmonica would be closer to the mark. This equates almost exactly to a shelf or two plus a couple of cupboards for your ingredients. You will also need your scales, thermometer and a measuring cylinder as well as some large mixing vessels and some form of hotplate. Your very own perfumery production and creativity centre has now been brought into existence (aka the back shed).

Of course you can wimp out and sell your essential oil neat and unadulterated, but it's nearly always better to have at least one product that will incorporate some of your oil. At the same time it will expand your market and add massively to your finances too. Decide on a single product first that you think will sell, then if that works stretch your wings and add a second and a third and then continue—forever onwards and upwards.

Why not start with a soap? No, not Corrie or East Enders: a genuine bar of soap, perfumed with your essential oil. There are several reasons why soaps are a good starting point: they are relatively easy to make, they will continuously showcase the fabulous aroma of your oil whilst in use—and at least

75% of the population know how they work, accept them and are comfortable using them. They also make excellent affordable presents that are easy to mail (unlike liquid products). Finally they should make a sizeable return on your outlay.

Soaps used to be made from animal fats, e.g. tallow and lard which contain fatty acids. These acids were then neutralised with alkali to produce the crude soap using a process known as saponification. Originally potash was used for the alkaline part but then other compounds were discovered and found to be much more satisfactory, such as the wonderfully named triethanolamine. However, since there are many excellent ready-to-use soap bases available, I strongly suggest that you use these first and then possibly produce your own soaps when you are confident with your production. It is certainly good fun to play soaps, but that is surely appropriate for a future date, especially when it's a wet playtime.

You can buy these ready-made soaps in 25kg tubs. They contain all the emollients, foam boosters and skin care ingredients necessary for a decent 1hour 45minute wallow. Some of the soaps also contain glycerine and are crystal clear, with the only things missing being the fragrance and colour. That is where you come in with your freshly brewed essential oil plus some dye.

The soap base will need to be melted and for that you will need some form of jacketed vessel similar to a cook's Bain Marie. This is because the melting needs to be performed gently—you can't simply blast it with a blowtorch or boil it up in the Christmas turkey roasting tin. It is probably worthwhile buying a Bain Marie from one of those companies who supply catering establishments and hotels: I'm sure that top hotel chefs don't muck about with pint-sized pots when producing sauces for a function of 300+ and their big boys' equipment is exactly what is needed now, if you know what I mean.

Once you have got your suitable vessel and its outer jacket is filled to the correct level with water, weigh out a decent amount—say 10kg or 20kg—of the soap. Cut it up so that it fits in the inner container, then put the vessel on the heat and allow the water to simmer. The soap will soon melt and should be gently stirred whilst keeping an eye on its temperature using your trusty thermometer. The best temperature to aim at is around 68ºC, so remove from the heat a few degrees before it reaches this temperature and then let it gently rise to the correct end point.

Once this temperature is achieved, carefully pour in your essential oil. Typically you will need a pretty small amount—a good starting point is 5ml of oil for every kilogram of soap. If you have started off with say 20kg soap then you will need 100ml of oil: this is poured into your measuring cylinder to the correct volume and then gently stirred in. Next add some water-based dye to tint it to an attractive pastel appearance. The dyes are also extremely concentrated and it is best to dilute them with water first and then add a known volume of this solution to your soap. That way you don't run the risk of producing lurid bars that would stand out in a Delhi street market.

Alternatively you can always go down the Organic route—or should that be root—and boil up various tubers, leaves and other goodies to make your colourings, such as 'Root of hemlock digg'd i' the dark, eye of newt and toe of frog' etc., etc. However, if you **do** want to go down the organic avenue, then that's up to you chum, but in my experience the result is likely to be 'Woe, woe and thrice woe.'

One of the obvious advantages of tinting your soaps is that you will then know exactly which soap is which (useful at stock-taking) and it leaves a bit of room for some artistic creativity.

"And what colour did you have in mind for your Lavender soap?"

"I see . . . . Mauve. Great. Really . . . inspired."

"And for the Jasmine soap? No, let me guess—I bet you chose a nice yellow. No, no-one told me, just a hunch—bet you were up all night dreaming that one up . . ."

Once your soap is fully mixed and is the right shade, let it cool slightly and pour it into soap moulds. These are available in all shapes and sizes, but good, inexpensive ones are available in PVC. As for sizes, the 100g mould is a very useful starting point although the 200g size is also popular. Using a jug, fill these moulds up to the brim with the molten soap or up to a given mark and allow them to cool naturally—you will get around 200 bars from your 20kg tubful, which will cover approximately a quarter of an acre of work top, so you might have to make the bars in 2 or 3 hits.

They will take about 3 hours to properly set and they must then be turned out of their moulds, trimmed to look neat and finally check-weighed to make sure that their weights are all within specification. Your soaps will then need wrapping and packaging to look professional and attractive. There are so many variables to this side of the market that I will simply say that today there is an increasingly critical view of overuse of non-essential packaging materials, so any form of excessive overpacking is likely to reduce rather than enhance sales. Simple cellophane wrappings and boxes will probably score in the long term.

A realistic retail price on a classy 100g bar of soap is around £3.50. Your soap starting material will cost around 20-25 pence. My case rests.

Another likely candidate as a simple vehicle for your star performer (i.e. your essential oil)—is shampoo. Again it is better to start with a standard unperfumed and uncoloured finished product and simply add the necessary fragrance and dye. The manufacture of shampoos from scratch is slightly complicated since you have to get the viscosity right and it is also trickier than for bar soaps to take a stock shampoo and to then perfume and tint it. But don't be put off—once you have got the method sorted, you can't go wrong.

The bottles and outer packaging will also need additional protection if they are to be sold via mail-order since we are dealing with liquids this time—or at least I hope we are. I can remember being a small child and washing my hair with some dry powdered shampoo concoction that had remained unsold for many years on a shelf of my Dad's newsagent shop. This powdered shampoo came in a sealed envelope and needed to be dissolved in hot water first. It resulted in a hot mugful of something akin to a caustic drain unblocker. It required 35 rinses to remove it from the scalp and if you were unfortunate enough to get it in your eyes, then God help you. In today's climate you would probably have called in the paramedics and had counselling. I know that my Dennis the Menace mug always tasted soapy no matter how many times I rinsed it and I would imagine this shampoo could have single-handedly wreaked havoc on any fish life for several miles downstream of the sewage works. Thank goodness we have progressed from there.

So, dealing with liquid shampoos, we now have to incorporate your essential oil. Unlike soaps, it is not always possible to simply add the oil and stir it in—sometimes it will remain on the surface and not become fully solubilised. On the other hand, it may go into the shampoo solution without any problem, but you will have to do a small-scale trial first.

There are ways to get a reluctant essential oil to solubilise into a shampoo and probably the easiest method is to mix it with—wait for it—a solubiliser. These are cosmetic raw materials designed to do exactly what it says on the jar: pre-mix the essential oil and solubilising agent together and then when fully mixed add the shampoo, a bit at a time until it is totally added and fully homogenous. If you think cookery—making gravy or custard without lumps—then you have already got the basics. The essential oil should remain in solution indefinitely and your shampoo will give off a fragrant waft every time you wash your hair.

Yes, I know I have simplified the process to a pretty ridiculous degree and you may find that making your own soaps and shampoos is a lot more complicated or problematical than the brief resumé outlined here, but the basic principles are exactly as outlined above. You may need to experiment with different solubilisers to find one that works best or changing the level of essential oil in your soap so that it actually does smell of something, or adding a little salt solution to your shampoo which amazingly increases the viscosity, or modifying your soap mould so that you can make a 150g submarine shape . . . . That's all part of the fun, all part of life's rich tapestry, all part of the challenge of overcoming the obstacles that magically appear, each with the sole aim of buggering things up as much as possible. With patience, you will win.

As for other products, there is no reason why you can't produce fragrant packs of pot-pourri or air sprays or bath oils or . . . For heaven's sake, when you have got a bottle of home brewed essential oil the world is your rather fragrant oyster (no, don't even think about it). You can even send the oil off to specialists who will blend it and then pack it in aerosol formulations. Rosemary air freshener. Lavender Shaving foam—a new first?! What about adding it to waxes to make all those addictively smelly candles—the ones that go straight into the 'present-recycling box', complete with donor info so you don't mess up and 'return to sender' in subsequent years? What about all the other sorts of perfumed waxes? What about . . .

I've been told to shut up.

Chapter 9

Just be Careful out there

Up to now, you may think I have been a trifle—how shall I put this . . . optimistic. Since going in for essential oils, it would seem that everything in the Lavender garden is coming up roses, smiling children play merrily in the flower meadows whilst the farmer, doing a fair Pop Larkin impersonation, trundles off to the bank in his Roller with a sack containing the days takings . . . . Perfick! I am sure someone has already commented in a style more in keeping with the utterances of the Royle family: "You can tell from page one that this joker's never done a bloody stroke of work on a farm!", whilst others may have given a rather more forthright critical analysis.

OK, time for a reality check. Yes, I can do doom'n'gloom, I can do all the negatives—in spades if need be. So let's look at all the pitfalls and the pratfalls. Let me spell it out in capitals: here is the **MESSAGE FROM THE DARK SIDE**. There certainly are some traps, ranging from the trivial, which will merely cost you time and/or money—to the serious, which may cost you time (inside) and/or money (lots). It's best that you know what they are and how to avoid them.

Firstly I would like to state the bleedin' obvious. You will need to determine any potential danger points that could possibly be found in the stillroom and any other areas where visitors may go. Hazard Analysis is the name of the game and writing up the wretched thing is what you will probably have to do. This means looking at every stage of your process and identifying each and every possible cock-up—sorry, hazard—that could occur. This covers events such as the water in the condenser accidentally getting turned off or what action you would need to take in the event of power cuts etc. Could someone trip and fall into your Florentine set? You will then need to write down what precautions are in place to make sure that nothing dangerous could possibly happen and you must also consider extremely unlikely but potentially dangerous occurrences e.g. what actions you would need to take if someone with the I.Q. of a Big Brother contestant were to turn up as part of a sightseeing party? Your local Environmental Health Office will be able to give advice, if you ask them nicely.

Yes, you must ensure that eejits can't access any areas where they might create mayhem. You must advise your insurers of what you are doing and tell them that you have put up all the necessary signs and warnings etc. and taken every conceivable precaution. All this is absolutely normal and what you would expect for an enterprise such as this in the 21st century.

Now the not-so-obvious. Put simply; don't **ever** consider distilling those 15 gallons of 1987 pea-pod wine inherited from Uncle Percy. For a start it's illegal and Customs and Excise officers would have a field day if they found out. They will almost certainly visit you at some point in any case (although you may not even know it) to make sure that you are doing what you are supposed to be doing: they are very astute and will spot anything remotely dodgy. In any case, when you distil wine to produce spirit, the first material that comes off is often contaminated with more volatile components (known as 'feints' in the whisky trade). One of the major components of this fore-run is methanol—unlike any schoolboy myth, this **does** make you go blind. Are you really certain of the point where you can change over from the mind destroying to the merely mind-blowing?

Neat alcohol vapour is extremely flammable, to put it mildly. It needs more cooling to condense it than water vapour—could you be sure that alcohol vapour would not come out of the condenser? I would hate to see the still take off like a Wallace & Gromit space rocket, especially if it takes out the stillroom roof at the same time. Facetiousness aside, it is an extremely dangerous thing to do, especially if you are not fully conversant with the process . . . and after all that, if your still has been previously used for distilling e.g. Lavender or Rosemary, what you would get out is likely to be totally undrinkable. What a bummer.

Up to now we have only considered the actual operation of the still and how it works: what about the oils themselves? Are they totally without risk? After all, they are completely natural, so there can't be anything remotely dangerous about them . . . . can there?

I have heard that argument many times from people who ought to know better and the obvious answer is "What about Hemlock, or Monkshood or any of the other 'natural' yet highly toxic plants that grow throughout the British

Isles. For that matter, what about the even deadlier plants that grow in other parts of the world—how about distilling some of them?" Yes, essential oils may be natural, but it doesn't follow that they must therefore be safe. In fact, playing around with a few of the really nasty essential oils is equivalent to what one of my American colleagues would call "opening a real large can of whoop-ass".

There are some essential oils that carry hazardous components that would affect everyone, whilst others may only cause a reaction with those who have, for example, sensitive skin. Others are known to cause skin sensitisation and you can suddenly develop a rash or reaction to something you've used many times before. It is certainly not as simple or straightforward as it looks.

I visited an old mate of mine a few years back who then ran a small unit dealing with perfumery items. He was cheerfully filtering an oil known as Litsea Cubeba from one tank to another and getting splashed for his troubles. This oil is a common component of many flavourings and perfumes and has a rather nice lemony aroma: that didn't stop the neat material from attacking the skin between his fingers. After a few months of working with this oil most of the skin on his hands had begun to split and crack and by the time I met him, the flakes were slowly extending all over his palms. He gave the impression he was rapidly morphing into a lizard, but I didn't tell him he was a prat, since he had managed to get his Doctor of Philosophy at just 24 (one of the youngest in Britain). It did however confirm that in most cases academic brilliance and common sense are mutually incompatible. What's wrong with wearing gloves?

Today it is recommended that all pure essential oils should be labelled with safety and usage tips. The International Federation of Aromatherapists (IFA) produces a list of 'cautionary essential oils' although there does not seem to be a definitive list of no-nos. Even some of our favourite oils that have been around for centuries are now no longer considered whiter-than-white: dear old Lavender oil (a Miss Marple favourite if ever there was one) has just been given the thumbs down in Canada along with another old timer, Citronella oil: they may no longer be used as a 'personal insect repellent, because of possible allergic reactions. The following text is taken from the Public Health Agency of Canada website:—

"Citronella and lavender: The re-evaluation of citronella-based insect repellents was completed in 2004. This re-evaluation was based on a limited amount of human health data that left a high degree of uncertainty about its safety. This uncertainty was incorporated into the human health risk assessments and subsequently the PMRA (Pest Management Regulatory Authority) was unable to conclude that insect repellents containing citronella were acceptable for continued use. Because of the uncertainties identified in the re-evaluation, the PMRA is proposing to phase out citronella-based insect repellents unless data to address the uncertainties in the human health risk assessment are generated and submitted by the manufacturers.

The manufacturers of the insect repellent containing lavender oil have decided to discontinue the product as a result of the re-evaluation. As such, the lavender oil products are being phased out in Canada by March 31, 2007".

However, before we all look skywards, roll our eyes and mutter, "The world has gone barking mad", I would simply like to mention Croton oil. It is one of only 3 oils banned under the 1968 Medicines Act—along with Chenepodium (American wormseed) and Savin oil. However, before about 1945 Croton oil would be found happily ensconced in British Pharmacopoeias and could therefore be dispensed as a 'purgative' or laxative by any doctor with a sadistic streak and a sense of humour. It was subsequently removed from medicinal use when it was discovered that "its action was deemed to be too drastic and violent for pharmaceutical purposes". Nice. It is now known to be intensely irritant and has been cited as causing tumours too. I know I shan't be taking it any more.

I had a three minute Googly trawl through the Internet on the subject "Essential oil + Safety"—this is a précis of a few comments from the first sites that popped up:

Bergamot oil is a potent photosensitiser, whilst Calamus oil is a potential carcinogen and is banned in cosmetics (presumably where 'calamity' comes from?) Cinnamon bark oil is an extremely powerful irritant and an even worse sensitiser. Copaiba can cause sensitisation reactions if it is old and oxidized and Inula graveolens is one of the most hazardous essential oils available. Pennyroyal Oil has for years been cited as an active abortifacient although there is now considerable disagreement over whether it can be considered culpable. Peru balsam is a very powerful sensitiser—RIFM (Research Institute for Fragrance Materials) recommend "not to be used as a fragrance ingredient". Rue oil is a terrible photosensitiser and sensitiser, with the recommendation—NEVER USE THIS

OIL—it is toxic and dangerous. Sassafras oil is restricted to such low levels in cosmetic products throughout Europe, that it effectively bans its use. Tests have shown it is possibly carcinogenic. Tagetes (Marigold) oil is a powerful photosensitiser—RIFM say a no-effect level is 0.05%. Therefore to use it on skin exposed to the light would be foolish. Tansy oil is extremely toxic, and of little if any use in aromatherapy. Verbena oil is an extremely powerful sensitiser—recommended by the RIFM "not for use as a fragrance ingredient". Massive percentages of adverse skin reactions are recorded from testing a whole range of verbena oils. Wormseed (Chenopodium) is extremely toxic and is banned from general sale in the UK because of the deaths reported from its consumption in the past . . . .

It may seem that most of these adverse reactions usually occur over a long period of repeated exposure, but some essential oils have active components that can give an immediate response. I would like to outline this salutary tale that I will remember all my life.

I was a young lad aged about 17 or so, working at my first proper job in the lab of an essential oil/perfume factory. Amongst the products the company used to make were Capsicum oil and Capsicum oleoresin, fiendishly hot but utterly natural products that were extracted from chilli peppers. They were used in the production of liniments and muscle-warming rubs where they were incorporated at fractions of 1%. The old boy who operated the plant making these Capsicum products stood about 5 ft 1 inches tall and always wore a grease-encrusted peaked cap, irrespective of the temperature. He had a protruding lower jaw that always seemed in need of a shave and was nicknamed 'Abdul' for reasons I can't divulge. One lunchtime I was sitting in the ancient Nissen hut that masqueraded as the works canteen along with some of the packing women and a couple of blokes, quietly playing cards and drinking my tea. Suddenly there was a commotion and looking out we could see Abdul racing towards the canteen. However, he shot past our door and flew through the next into the kitchen where Nancy used to make the dinners and brew the tea. We all stared through the huge serving hatch in disbelief, as first poor old Nancy was shoved out of the way, closely followed by all the washing up that was stacked up on the draining board. To the sound of smashing cups and plates, the taps on the sink were turned on full as Abdul frantically climbed out of his wellies and his dungarees. He finally hurled an unsavoury pair of pants to the floor and propelled himself buttocks-first into the sink, although we all noticed that he still kept his cap on.

I hadn't worked at the factory for very long, so I enquired if this was typical behaviour, and should we all join in? I was rather hoping it wasn't compulsory.

I eventually discovered that Abdul had apparently got some Capsicum on his hands: since they had the appearance of being constructed from recycled tractor tyres, he hadn't noticed this contamination—until he went to the gents . . . .

Who says natural products are harmless?

I could go on . . . but all that would happen is that you would probably say "Well, if it's all so dangerous, why is this dolt writing all this stuff, encouraging us to get a Still and then telling us they produce lethal oils?" This is followed by the sound of a small book being hurled into the nearest bin at Mach 2.

Stop! It's **not** all poisonous sensitization and inevitable sprouting tumours—most essential oils are safe, innocuous little varmints that wouldn't hurt a fly. Alright, many of them **do** kill flies, but they are still as safe as houses. No, honestly, they really are. It is simply that essential oils smell—that is their job—so they must therefore contain odorous chemicals.

After my first job, I then spent many years doing organic research into essential oils. It sounds glamorous, but looking at all the 'ingredients' of a single essential oil was probably the most boring work imaginable. The problems arise because every oil contains hundreds of component parts: some of these are very complex or are present in only microscopic amounts and have yet to be synthesised whilst others are simple bog standard organic chemicals, common as muck and used in every-day products. Some smell nice whilst others make you gag, some smell like the essential oil they come from whilst others smell nothing like—or like nothing. The light volatile components found in the oil that is distilled first contain the 'top-notes' so beloved by the perfumers, whilst the later distillate contains the less volatile and usually larger molecules that supply the heavier 'base-notes'. In theory, if you were to re-mix all the hundreds of components back together in their original proportions you would end up with an oil that smells identical to the original natural material.

If only it were that simple . . .

In an essential oil, amongst this whole seething organic population there live some bad boys—chemical Kray Brothers if you like. Usually they are there in minute quantities and can be ignored, but in some cases there are enough baddies present to make the oil suspect or even a no-go area. So how do we know which oils are safe? Well, if you stick to oils that are currently available and freely on sale then you can't go too far wrong. Aromatherapy has been using essential oils for millennia and knows which ones to avoid: you can always check up the current status on the Internet.

Finally if you want to try something unusual (e.g. that weird weed with the jagged leaves that is growing behind Michael's bedroom) then at least send the oil off for chemical analysis before you even think of using it. You never know—it may be toxic, hallucinogenic or even emetic . . . but it might, just might, be worth a mint. At least you will know if it is OK to leave it on your skin and once it has been given the green light you can use it in your products with a clear conscience.

Chapter 10

And so to Bed ...

If this booklet has tweaked your interest in the production of essential oils, then I suppose it's all rather splendid—and I'm very pleased that you are happy too. No, really, I am. You may, however, think that a lot of the items discussed in the preceding Chapters may be hard to source: you are certainly not wrong there. Even with the power of the Internet, with its all-powerful Search Engines and other devices far too nerdy to ever contemplate, it is still hard to track down many of the items required for Essential Oil distillation.

To start with, if you type "Essential Oil Still" into a typical search engine, then quite a few companies will crop up, with a surprisingly large proportion based in California (where else?) They all seem to be offering similar, rather interesting contraptions: the larger stills look suspiciously like Stainless dustbin/Dalek combos with propane burners, whilst the smaller ones are made out of beaten copper and look very pretty with lots of curly dangly bits. These smaller ones have either gas burners, a 110 volt electric element, or are designed to sit on a kitchen range, like a do-it-yourself Chernobyl kit.

Read and believe. "Distil your own pure essential oils and hydrosols with a beautiful handmade copper alambic (sic) still". "Experience the art and science of true aromatherapy distillation". "The act of distillation is the art of the invisible becoming visible". "Aromatherapy may be defined as the use of pure essential oils—their scent—from plants for physical and mental well-being".

There is loads more tosh where this came from, with most of it written in the same pseudo—mystic vein. Fine if you are thinking of setting up your still in a shop in Glastonbury High Street, otherwise stick to 'Crystals for Feng Shui' or 'Witchcraft & Pagan Supplies'. Cynical? Moi?

Back to our Still suppliers. In addition to the above, there are several glass laboratory-scale stills on offer 'for the home distiller', with flask capacities ranging from 2 up to 50 liters—sorry, litres. None of the above is really suitable for commercial production but they may prove useful for checking out your plant materials prior to production. That is, if you find a way of actually getting hold of one.

The surprising fact is that most of the American companies will only ship 'UPS, Ground only' which I presume means 'Tough bananas if you live in Hawaii'. As for exporting; it doesn't appear to enter their mindset at all, although some may grudgingly allow their goods to sneak into Canada. When I telephoned one Californian supplier, their rep responded to my export enquiry with an audibly large intake of air through the teeth: "I'm sorry sir . . . we just can't take the risks that would arise if we were to send them abroad. Hell!—With all our local aggravation, it's bad enough selling them within the State . . ."

Alan Sugar, eat your heart out.

Another interesting still with a capacity of 25 litres is available from New Zealand. It gives the following information on their quaintly dyslexic website: "In New Zealand and other certain countries, it is legal to also use this unit for the home distillation of alcohol. In many countries this is not the case. Please check your local regulations . . ."

It has always been a great disappointment to me that since I was born, I have lived in a place that has come under the category 'Many Countries' where it is 'Not The Case'—i.e. where the home distillation of alcohol is taboo. In fact I would go further than that: it is viewed by the UK authorities as a far more heinous offence than, say, exporting enriched uranium to wildly unstable countries or packing nail-scissors in your hand luggage. I don't need to check my local regulations: I just know they will go utterly ape if I even think about asking if I can have a go.

The data sheet then goes on to say—"To avoid the risk of boiler implosion (boiler walls being sucked in) ensure the end of the outlet hose from the condenser remains above liquid level at all times." In the circumstances, you probably will.

Yes, I know what readers are thinking: "I can see where he's coming from. He is dissing everyone else's stuff first: then he is going to introduce his own equipment and say how wonderful it is and try and flog it. Devious or what?"

Correct! I would also like to compliment you on your astute powers of deduction.

Yes, my little company does indeed make stainless steel stills and supplies them complete with all the Florentine glassware. They are a bit tasty. You can find them on the website www.gnltd.co.uk where you click on the link to 'Distillation Equipment'.

Alternatively you can telephone + 44 1233 770780 and ask for more information. I had better say no more on the subject or I will lose the remaining fragments of credibility I ever thought I had.

For second-hand laboratory equipment I recommend that you try Severn Sales, 1 Lodge Road, Kingsway, Bristol BS1 5LD. Tel: 0117 960 8858—a company that I have always found to be remarkably helpful. Even better, although they have a web-site, you actually speak to a human being who finds out what you want and then tries to assist. They can also supply simple chemicals like the anhydrous sodium sulphate that is needed for drying your oils.

Another option is dear old eBay—you will be amazed at the number of disenchanted amateur boffins there are in the outside world, ready to hang up their test-tubes, or whatever it is they do. The price you will pay for their equipment is usually a fraction of the cost of the new item.

You will be surprised just how expensive new laboratory kit is: if you want to find out the real cost, pour yourself a stiff one and then contact any of the laboratory supply companies that are listed on the Internet e.g. Fisher Scientific—www.fisher.co.uk. or York Glassware—www.ygs.net—to name just a couple. They usually all do basic chemical supplies as well as mantles and stands.

For Soap making sundries, I would suggest 'Soap Basics' The Cottage, Station Hill, Chudleigh, Newton Abbot, Devon TQ13 0EE. Tel: 01626 854577. Their website is www.soapbasics.co.uk and in it they quote:—

". . . the UK's leading online retailer of soap making supplies. Our range of products includes soap moulds, soap bases, melt and pour supplies, soap making instructions, mica powders, liquid dyes and soap packaging". This company comes highly recommended and receives plaudits in particular for their soap moulds.

There are plenty of others who can assist:

The Soap Tub (www.meltandpoursupplies.com),

The Soap Kitchen (www.soapkitchenonline.co.uk),

Just a Soap (www.justasoap.co.uk)—and there are more bubbling away where that came from.

I won't give details of any packaging suppliers since there are so many of them in every locality it would merely be superfluous. In any case, you are just as capable as I am of looking in Yellow Pages or on the Internet. What do you want—spoonfeeding? Oh, for goodness sake—well, all right then, seeing as you have bought the book. Here are the first three out of the hat: for nice glass bottles try Continental Bottle Co—Tel: 01606 862525, or Wains of Tunbridge Wells—Tel: 01892 521666 or Nekem Ltd.—Tel: 01428 683030. What about nice little boxes? O.K. Try Boxmart—Tel: 01543 411574 or A C Cartons—Tel: 01354 741884. You want to know where to get some nice ribbons?

Oh for Pity's Sake! This is positively the last one; you're on your own now. To start with, contact Partybox Ltd.—Tel: 01483 486000 who should be able to help.

Finally and perhaps most importantly, what about the plants you should be growing? After all, this is what your enterprise will depend upon—the growth of healthy crops that contain their full complement of essential oils. You don't have to just grow Lavender—there are loads of exciting possibilities awaiting discovery. The logical way to tackle this selection would be to take advice from as many experts as possible.

In my own part of the United Kingdom there are several agricultural colleges: these are the haunt of a veritable bevy of botanical boffins and boffinettes who are usually extremely helpful in coming up with answers to your dumb questions. Be persistent. Be tenacious. If all else fails, be wheedling/aggressive/pleading/pitiful or whatever it takes to get a serious answer to your question: "I want to grow plants for the production of essential oils. My soil is clay/sand/peat/landfill and I have xx hectares/acres/rods, poles or perches of land to work on. Any suggestions?" Sooner or later, someone will be tempted to flex his/her brainpower and spend some time cloistered in front of a computer screen or ensconced in the library. Hopefully an answer will be forthcoming—which, with any luck will be a list of botanic candidates suitable for growing in your area.

Check out the local nurseries too and pick their brains. If they have similar soil to yours and they are currently growing Rosemary for example, and it is 'vigorously thriving' (to use a Titchmarshism), then take this info on board and add it to your notebook.

Snoop. Be nosy. It's not illegal but it can be very instructive, especially as gardeners just can't help themselves. They seem to be pre-programmed to follow an irresistible urge: it directs them to unquestioningly give useful tips to any passer-by that happens to glance over their hedge. They just **have** to impart horticultural profundities—just in case you were wondering why their Lavender was growing so well. They positively glow with pride if you dare to offer the slightest praise regarding their Lemon Balm. Cuttings will be eagerly thrust upon you, whilst Latin names are tossed about like confetti. If you're not careful, tea will be offered and invitations proffered for further visits. Any initial antagonism as to why you were trespassing in their garden in the first place will be quickly dispelled: "No, don't worry: it happens all the time. Only last week we spotted someone crossing the lawn to admire Harold's red-hot pokers. Apparently she couldn't resist . . ."

Add these nuggets to your 'Book of Wisdom' too, then study, digest and finally take the plunge and get planting.

That's it. You can now set off on your great essential oil adventure/journey/catastrophe. Good luck.

I've been told the light is going off since it's well past midnight and we've got an early sta

Compulsory Thanks

Books and other publications of this type normally have a section in which the author acknowledges and thanks all the contacts, friends, acquaintances and even enemies who have assisted in '*making it all possible*'. Maybe I'm a bit cynical and I probably ought to feel a bit guilty about it, but usually I skip over this piece of fawning gratitude, dedicated to people I don't know, I probably never will know and certainly never would want to know. Honestly—how many saintly people **are** there in this world, prepared to selflessly proofread all 18 chapters with only a Tilly lamp for illumination, or take the author on a 360 mile detour upriver to collect his malaria tablets? Talk about making you feel inadequate.

In the late 60's and 70's I was a member of a rather mediocre pop group that regularly did bookings on Friday and Saturday nights. I can still remember playing at some of the wedding receptions where we appeared as 'The Evening's Entertainment'. Most were spectacularly tacky affairs in village halls and usually ended up with feuds between rival clans, complete with sobbing mothers and threatening fathers and concluded with a fight in the car park. On one glorious occasion we even enjoyed the spectacle of projectile vomiting from a drunken fifteen-year-old sibling who scored a perfect bull's-eye on his sister's back—who just happened to be the bride. However, what all these functions had in common was the 'Thank You' speeches. "Thank you to everyone for coming." "Thank you to the bridesmaids—you all looked really smashing!" "Thank you to Auntie Brenda for putting on such a lovely spread" and "Thank you everyone for helping to clear up".

It seems to be an absolute golden rule that we must always show our appreciation, express our thanks and acknowledge our colleagues, friends and relatives. In fact it is a rare event indeed to find someone brave enough to ignore protocol and **not** give thanks to anyone or say a friendly hello.

Many years ago I was driving home in my Renault 4 and so bored that I was reduced to slightly listening to the only noise my car radio would emit i.e. the local radio station. It was broadcasting 'Live and direct from your very own Kent County Show' and the DJ/presenter was rabbiting away in his mid-Atlantic accent, desperately trying to be slick and make it sound exciting.

"Now we are going to open up the marquee so that all you people can come on in! Then we'll play **your** dedications **live** to the people of Kent!" "And first in line is David. Where are you from David?" David gave the name of a village and I immediately came off autopilot and pricked up my ears. The place David mentioned comprised of a few houses plus a pub, but also an absolutely vast psychiatric hospital that seemed to cover an area the size of Luxembourg. In fact, the village was synonymous with the hospital, so when the presenter asked "Is that in Kent, David?" I realised that he didn't know what everyone else in Kent knew. I suspected that this interview may suddenly become interesting, especially when the presenter asked, "So, Dave, have you

Making Sense of Making Scents

had a good look around the show?" and received a loud surly reply "No! . . . Bin Nowhere!"

I had to take my hat off to him: the presenter didn't seem at all fazed and there wasn't even a trace of panic in his voice. He merely chuckled and said "So, you've ignored the rest and come straight to the best—the Radio Kent Marquee! Wise move there, Dave! And have you any dedications to any of your friends out there?" There was a full ten seconds of silence whilst I held my breath. Then Dave answered—and he certainly didn't let me down. It was clear and extremely loud: "NO!! I don't want to say NUFFING!!" Another few seconds of total silence passed. Then an even louder: "To NO-ONE!!"

"OK, er, let's continue with the show and move straight on with Herman's Hermits". In the background you could just make out the sound of frantic scrabbling as the security men ejected Dave. I had to stop the car at least four times on the way home.

Well, I certainly **do** want to say something, but I promise to keep these thanks remarkably short. I would like to thank the people at Givaudan, aka Quest International, aka Proprietary Perfumes Ltd. in Ashford, one of the U.K.'s major perfumery companies, where I used to work many years ago. They kindly allowed me full access to their wonderful library of perfumery books that go back to the early part of the 20th Century: many of the photographs used in this book originate from this source. Their chief librarian, Richard Butcher, couldn't have been more helpful and makes a mean cup of coffee. Other colleagues from Quest who assisted included my old boss, the late Dr. John

Belsten who helped by checking information at the library (bless him!) as well as my old mate Ian Payne who also came up trumps.

Anne Welby from Welby Healthcare Ltd. up in Bothwell, Lanarkshire is Queen of Aromatherapy and very kindly wrote a complete background on the subject and brought me up to speed.

I must also mention Peter Willoughby of Willweld Alloys Ltd. who is both my colleague and my factory neighbour. He manufactures my Stills—painstakingly slowly, but I must admit, beautifully too. His stainless steel fabrication is sublime and welding cognoscenti have been known to discreetly run their fingers over his seams in awe when they think no one is looking.

A large helping of gratitude must be passed on to all the people who have bought Stills from me over the years, although one in particular must be mentioned. Mike Hicks lives on the remotest of the Scilly Islands (St. Agnes) and farms in a situation as near to paradise as is theoretically possible. His tiny fields are typical of all the Scilly Islands and with the balmy climate and the crystal blue sea lapping at their edges, they are perfect for planting a whole spectrum of wonderful aromatic plants. His unselfish input has been immense and I can only say how glad I am that he has made a success of his essential oil venture. This is at a time when the traditional Scillonian crops (daffodils and variations thereof) are being squeezed out of existence due to the fact that tons of wretched odourless African flowers can be imported by the jumbo-jetful and sold at a price less than what it costs the locals to get their flowers from the Scilly Isles to Penzance.

Thank you, Mike.

Another set of customers I have to mention is the happy band who make up Carshalton Lavender. They valiantly fought off the building developers who coveted their tiny acreage of prime allotments close to London and after winning their battle turned it into a successful Lavender plantation. They now distil their oil on site and have extremely popular Open Days which allow the public to watch the distillation, buy products and stand in a sensuous sea of purple whilst being totally surrounded by housing! My main contact there is Laurie and when we correspond, I am always styled as 'Dr Bunsen Honeydew', whilst my responses always go to 'Beaker'. In fact the photograph on the front cover is one of Beaker's and shows just what can be achieved in an industrial setting . . . Thanks Beaker.

I would also like to express my gratitude to David Burnett for encouraging me to write this book in the first place. He had seen an article of mine on Stills and Essential Oils in 'The Smallholder' magazine and had made the quantum leap to see a book hidden away in it somewhere. He also entrusted me with his copy of 'The Still Room'—a wondrous rare Victorian booklet that I mentioned in Chapter One. It is currently doing sterling service as my beer mat. Only joking, David. Honest.

Finally and above all others, I would like to thank my wife Margie for helping make this book what it is. Her excellent hawkeyed proof reading spotted hundreds of mistakes, duplications, repetitions etc. and all the expletives were carefully excised. She generally helped me construct sentences so that they went above pre-school level and also let me know what parts of my writing could be interpreted as offensive, racist, bigoted or psychologically harmful. (**Not that he took any notice—Margie**) Still, she got her own back after being woken for the third night running at 2am to inspect my latest outpouring: I think Chapter 7 was re-written five times.

Even my kids joined in with the occasional "Daaaad. You can't say **that!**" although on the whole they were touchingly encouraging.

Thanks to you all . . . and to anyone else who knows me.

Printed in Great Britain
by Amazon